An archaeology of innovation

MANCHESTER
1824

Manchester University Press

Social Archaeology and Material Worlds

Founding editors
Joshua Pollard and Duncan Sayer

Series editors
Chantal Conneller, Laura McAtackney, and Joshua Pollard

Social Archaeology and Material Worlds aims to forefront dynamic and cutting-edge social approaches to archaeology. It brings together volumes about past people, social and material relations, and landscape as explored through an archaeological lens. Topics covered may include memory, performance, identity, gender, life course, communities, materiality, landscape, and archaeological politics and ethnography. The temporal scope runs from prehistory to the recent past, while the series' geographical scope is global. Books in this series bring innovative, interpretive approaches to important social questions within archaeology. Interdisciplinary methods which use up-to-date science, history, or both, in combination with good theoretical insight, are encouraged. The series aims to publish research monographs and well-focused edited volumes that explore dynamic and complex questions, the why, how, and who of archaeological research.

Previously published

Images in the making: Art, process, archaeology
Ing-Marie Back Danielsson and Andrew Meirion Jones (eds)

Neolithic cave burials: Agency, structure and environment
Rick Peterson

The Irish tower house: Society, economy and environment, c. 1300–1650
Victoria L. McAlister

An archaeology of lunacy: Managing madness in early nineteenth-century asylums
Katherine Fennelly

Communities and knowledge production in archaeology
Julia Roberts, Kathleen Sheppard, Jonathan Trigg, and Ulf Hansson (eds)

Early Anglo-Saxon cemeteries: Kinship, community and mortuary space
Duncan Sayer

An archaeology of innovation

Approaching social and technological change in human society

Catherine J. Frieman

Manchester University Press

The right of Catherine J. Frieman to be identified as the author of this work has been asserted by her in accordance with the Copyright, Designs and Patents Act 1988.

Published by Manchester University Press
Oxford Road, Manchester M13 9PL
www.manchesteruniversitypress.co.uk

British Library Cataloguing-in-Publication Data is available

ISBN 978 1 5261 3264 2 hardback
ISBN 978 1 5261 7178 8 paperback

First published by Manchester University Press in hardback 2021

This edition published 2023

The publisher has no responsibility for the persistence or accuracy of URLs for any external or third-party internet websites referred to in this book, and does not guarantee that any content on such websites is, or will remain, accurate or appropriate.

Typeset by Newgen Publishing UK

Contents

Figures

Tables

Acknowledgments

I wrote most of this book on unceded Ngunnawal and Ngambri land – beautiful Country where I have been privileged to live for the last nine years. I acknowledge the Traditional Owners of this land and pay my respects to their elders, past and present.

This book, like many books, had many sources of support and even more midwives. The majority of the manuscript was written as part of my Australian Research Council-funded DECRA project "Conservatism as a dynamic response to social and technological change" (DE170100464). Chapter 5 is largely the result of research carried out as part of the Australian Research Council discovery project "Beyond migration and diffusion" (DP160100811). I first drafted Chapter 1 during my time as a visiting senior fellow at Topoi/DAI-Eurasien Abteilung in Berlin and as a visiting scholar at the University of Durham. I thank Florian Klimscha, Svend Handsen, and Ben Roberts for arranging these visits. Large parts of Chapter 3 were written while I was a Mercator Fellow at the University of Kiel and Landesmuseum Schloss Gottorf. Thanks to Berit Valentin Eriksen for her support. Some of the material on fragmented assemblages in Chapter 2 was first written (but never used) during my time as a post-doctoral research fellow at the Research Laboratory for Archaeology and the History of Art, University of Oxford, where I worked with Peter Bray and Mark Pollard to develop the idea of archaeo-prosopography as part of a project supported by the John Fell Fund and the Prehistoric Society.

I am grateful to many other people as well. I started thinking about re-framing innovation studies with archaeological literature while talking about my then in-progress doctoral research with Peter Bray and Duncan Garrow, both at Oxford at that time. Pete also provided me (albeit unknowingly) with the "bad anthropologists" line in the introduction. Further ideas were developed in conversation with my archaeology colleagues Florian Klimscha, Shinya Shoda, and Ben Roberts. A number of people read part or all of the text as it developed. Special thanks for this are due to my colleague and friend Phil Piper, who read the whole manuscript, and parts of it more than once. Guillaume Molle not only offered valuable critique on the manuscript, but also kindly provided two beautiful drawings for the book. Others who deserve my gratitude for their critical feedback and wise suggestions on elements of the text include Jenny Davis (who also generously provided me with then-unpublished chapters of her own monograph at a critical juncture), Francesca Merlin, Robin Skeates, Shinya Shoda, Tim Denham, Matthew Spriggs, Mathieu LeClerc, Jayne Wilkins (whom I thank in particular for nudging me to consider the innovativeness of Middle Stone Age African hominin populations), Miljana Radivojević (who generously provided me with chapters of her unpublished Ph.D. to consult), and Frances Morphy (who offered valuable critical feedback on the original grant application – some of the text of which is scattered within this manuscript).

Special thanks to Sally K. May for allowing me to adapt material from our ongoing collaborative work exploring the social aspects of Aboriginal Australian rock art and to use it as the case study in Chapter 4. Thanks to Ronald Lamilami for permission to study the rock art from Djulirri.

A number of people allowed me to use their photographs and figures. Thanks to Tim Maloney for his beautiful photograph of a clear-glass Kimberley point that graces the cover of this book. Thanks to Stuart Bedford for his photograph of Lapita pottery. Thanks to Lindsey Ems for her photograph of an "Amish mobile phone." Thanks to the European Association of Archaeology for permission to reproduce Figure 5.3 from Andrew Sherratt's 1993 paper in the *European Journal of Archaeology* (Sherratt 1993). Thanks to James Flexner for allowing me to use his lovely basemap to make my own map of the Lapita world.

Other contributions were less tangible, but just as important. Conversations with Carly Schuster, Simone Dennis, Mercedes Okumura, Matt Walsh, and Robin Torrence challenged me to think deeper and work harder to support my positions. Matt, in particular, made me think more deeply about evolutionary approaches and sent me considerable amounts of edifying reading material. Tim Maloney gave me a number of valuable references for research into Pleistocene Australia. Two archaeologists I've never met, Patricia Crown and Barbara Mills, both very kindly shared with me PDFs of their hard-to-find-in-Australia research. James Flexner, Matt Pope, Martin Porr, Miljana Radivojevic, Ben Roberts, and Ben Marwick all helped me acquire hard to find articles and book chapters. Huge thanks to #ArchaeologyTwitter and #AcademicTwitter for helping source the really obscure stuff – the number of people I've never met who were willing to take time and track down an article, photograph an old magazine, or otherwise find me a reference was simply astounding. They also regularly laugh at my jokes, so that's pretty helpful too.

Thanks are also due to Meredith Carroll and the Manchester University Press team for their support (and gentle nudges as deadlines approached/passed). The proposal and manuscript peer reviews were among the most useful, insightful, and collegial I have ever received; and I am grateful to Manchester University Press for arranging such a stellar process and to the reviewers for their time and attention. Nigel Jonas helped produce the index.

This book is dedicated to Ash Lenton. Back in 2008 or 2009, well before I finished my doctorate, I went home after a frustrating supervision and told him I was going to sort out innovation because no one else was doing it right, and he didn't laugh at me. His support, feedback, good humor, and forbearance made this book possible.

Abbreviations

BCE	before the common era (calibrated)
bp	before the present (non-calibrated)
cal BP	before the present (calibrated)
CE	common era (calibrated)
ECT	evolution culture theory
NCT	niche construction theory
RIS	regional innovation system
SCOT	social construction of technology
UTAUT	unified theory of acceptance and use of technology

Introduction: Loomings

How do we proceed with an archaeology of innovation?

Any book on innovation worth the value of the paper it's printed on should probably start with a few sentences on how exactly it understands innovation, what it defines innovation to be, and what roles it sees innovation playing in society. Innovation is a slippery term. It is culturally loaded, and has been since it entered common usage in English in the post-medieval era; but its meaning has shifted notably over time and, unsurprisingly, its usage continues to alter from one context to the next. Innovation in its most general sense can be understood as any new thing, idea, or practice taken up by a person or people – this is the definition put forward nearly seventy years ago by the anthropologist Homer Barnett (1953, 7). More materially minded writers, whose interests tend to lie in economics or technology studies, treat innovation as a synonym for the development of novel techniques, technologies, or organizational principles; while others separate the actual creation of the new thing or idea (its *invention*) from its dissemination and uptake (as an *innovation*) (Godin 2015b, 24–5). In this book, *innovation* should be understood as a novel thing (tangible or intangible) or combination of already known things into a new formation that was adopted widely enough to be archaeologically visible, and also as the slow and punctuated process of its adoption.

Moreover, within this definition, I want to acknowledge clearly that innovation is not a neutral term: it has weight and cultural baggage that are rarely acknowledged in academic writing on technological change. We, in the twenty-first-century Anglo-sphere, live in a predominantly innovation-positive culture. Innovation itself carries positive connotations, and a resistance to innovation or a preference not to adopt any or most innovations carries negative ones. These ideas have historic weight as they spring from particular strands of Enlightenment thinking about technology, society, and human worth that are tightly bound up in both the nineteenth-century European technological efflorescence (usually referred to colloquially, and not without value judgments attached, as the "Industrial Revolution"), and the concomitant European colonial expansion in which the technological dominance of Europeans played a major role in their ability to conquer. Yet, even though the term is ambiguous and carries enormous, perhaps insurmountable, implicit associations, I have chosen to retain it in place of more neutral terminology, such as technological change, which is preferred by some of my colleagues (e.g. Schiffer 2011).

Bricolage as meaning-making

When I started to write this book several years ago, I imagined I would write a book about innovation for archaeologists in which I distilled insights from a range of disciplines to help my archaeological colleagues grapple with the way people shape their technology and society. As I started to read and research – and, most critically, write – I found that archaeology actually has quite a lot to offer all those other disciplines too. Not only do we have millennia of case studies and examples of how ideas develop, new things emerge, and others disappear, but we also have a unique interpretative process. Archaeologists often lament that there is little to no purely archaeological theory, but that we continue to draw on insights from outside our discipline, particularly from anthropology, sociology, and philosophy, in order to understand the past. There is often a sense of inferiority attached to this observation: that is, that if we were a serious or proper social science, we would have developed our own body of uniquely archaeological theory rather than being forced to look outside our field for inspiration. I can't agree with this attitude. To quote a former colleague and old friend who regularly debated theory with me in a variety of Oxford pubs, "we have to stop thinking of ourselves as bad anthropologists." Instead, I think it is worth looking at the strengths of the archaeological approach – an approach that I utilize in all its glory in this book.

I have spent much of the last decade teaching the history and theory of archaeology to undergraduates. In every cohort, there's usually one who objects to reading articles by historians or sociologists as part of their degree. "What does this have to do with archaeology?" they ask. "Why should I have to read about feminism [or structuralism or actor-network theory or art history] in order to study archaeological sites and materials?" What I have tried to teach my students (with greater success than the occasional vocal complaint might suggest) is that interpretative archaeology is much like the more tangible bits of archaeology. Just as we construct a plausible pot from small fragments, experience, and guess work, we apply the same sort of creative *bricolage* to the wide body of social theory literature on which we draw. We juxtapose clever ideas developed by theoreticians from across a spectrum of disciplines with (inevitably fragmented) archaeological data, scientific analyses, and historic interpretations in order to construct new and different visions of past worlds and the people who inhabited and created them. Because the people in the past are intangible, it is up to us to animate them, and we do this by trialing a bit of this and a bit of that to see what makes the most sense. I have been told by colleagues in anthropology and sociology that this approach is deeply unsettling. Our refusal to situate ourselves in a single body of theory and our willingness to decontextualize ideas developed within specific, present-day social contexts makes them uncomfortable and liable to question the results of our analyses. But archaeology is a magpie of a discipline. Aside from digging trenches (and there's a good case that we can attribute these skills to the military training of a number of early archaeologists), our methods have always been drawn from a number of fields; and our discipline operates by picking up new and shiny methodologies and applying them more or less successfully. Archaeological scientists in recent years have borrowed genetic sequencing, isotope studies, and ever more complex ways of assessing the dates of ancient materials and sites from biology, chemistry, geology, and physics. This creative recombination and appropriation of methods and ideas – the *bricolage* approach, as I have termed it above (with a nod to Lévi-Strauss 1966) – is one of the strengths of archaeological interpretation (and perhaps one you'll keep in mind while you read Chapter 4). Faced with the impossible task of reconstructing lost worlds from bits of rock and broken crockery, we have drawn on all possible sources and written libraries' worth of books, articles, and site reports.

In the chapters that follow, I jump from economics to organizational studies to sociology to anthropology to the history of technology and back again. I weave the variety of approaches to innovation developed by scholars in these disciplines together with archaeological case studies; and, in so doing, I have found myself not just writing a book on innovation for archaeologists, but actually using this intensely archaeological approach to study and critique the concept of innovation itself and the social and technological assumptions that underlie it. Archaeologists often complain about "archaeology" being used as a metaphor in other disciplines. We always get a good giggle at archaeologies of literature and advertisements for professorships in galactic archaeology, but, in this case, I think it applies. I have treated the elements of innovation and the various approaches developed in a range of fields to study them as I would treat a fragmented artefact assemblage. I have laid them out, put them in order, rearranged them until their edges seem to match up, and glued the resulting hodgepodge together into a shape that makes sense to me (and hopefully also to you too, Reader). At times – such as when I managed to connect Schumpeter to Strathern via Tech Bro disruption – I have wondered if I was perhaps taking the approach too far, but I think the results speak for themselves. Although I know that most of the people reading this book will be archaeologists (it is the sad nature of disciplinary silos that we are often under-exposed to relevant materials outside our specific fields), I hope it is legible and comprehensible to any brave anthropologists, sociologists, or historians of technology who might pick it up. To those readers, I say, please work through your discomfort: our method may seem mad, but it serves to confront our own assumptions, forcing us to ask why certain interpretations seem obvious or common sense, whilst others are wildly implausible. We developed it because the past is dead and gone, and this academic necromancy allows us to reanimate and interrogate those minimal, corporeal traces that remain.

Approaching social and technological innovation

To make my approach clear, this book stems from my profound skepticism about the just-so stories we tell and are told about how technological change works. These stories implicitly support the muscular technological development of the modern western world – no doubt because they are typically written by western scholars who focus on the last 50, 100, or, if they are feeling very frisky, 200 years of social and technological change. The achievements of important people (and, let's be honest, typically these people are white men) feature prominently, often at the expense of their assistants, informants, spouses, families, mentors, and collaborators. Moreover, these stories frequently equate change with progress and rarely question the latter. Archaeologists too use this post-Industrial-Revolution cognitive framework, even though much of the material with which we work pre-dates it; and we know – both from historical writing and from anthropology and ethnohistory – that the small-scale societies that characterized the pre-modern era often operated with profoundly different internal logics than our own.

The ideas that I develop here began to emerge during my doctoral research at the University of Oxford. I set out to discuss the shift from Stone Age to Metal Age in Europe by examining a number of different stone tools that are widely thought to be copies of metal (Frieman 2012c). What I found in the course of this research is that, inevitably, the explanation for the emergence of these complex lithic forms was much more complicated, and varied considerably by region, period, and local social context. The big revelation for me, though, was that for 200 years archaeologists had just blindly accepted, almost without question, the idea that metal was so obviously a step forward that prehistoric people would have gone to great lengths to shape stone in its image. This is a profound misconception of how people engage with, learn about, and adopt new technologies – not to mention an under-estimation of the skills and experience needed to make elaborate lithic objects. Yet, this narrative of stone being neatly replaced by copper or bronze after a period of covetous imitation persists. (And you'd better believe that, by now, I have published enough papers critiquing this framing that I am constantly frustrated by its reappearance! I've tried to make this point again in Chapter 3; maybe it will stick this time.)

Although some archaeologists might bristle to be told so, this model of technological replacement via desirous imitation is profoundly complicit in colonialist patterns of thought, just as a focus on the value and significance of technological (and social) change (rather than persistence or maintenance) is innately Eurocentric and masculine (Montón-Subías and Hernando-Gonzalo 2018). Wherever possible, I have tried to lean on the work of indigenous, post-colonial, and feminist scholars in order to challenge these narratives. Moreover, I have done my best to lay out the implicit assumptions within the wider bodies of literature I cite, and I have used archaeological material and the *bricolage* approach in building these critiques.

Over the following chapters, I use this unique perspective to develop an understanding of innovation in human society that is not tethered in the economic structures of the modern world. To do so, I tack back and forth in time, exploring case studies from our earliest hominin ancestors through to the early twentieth century, and contexts from the Pacific to Europe to Australia and the Americas. I examine all sides of the innovation process, from why innovations develop and spread, to how innovations are communicated, to why some innovations fail, and why certain groups of people seem particularly predisposed to innovate. In the end, I propose a social model of innovation, applicable not just in studies of the past, but to innovation in the present as well. Since this book is in many ways an excavation of an idea, I have framed it around a quest for knowledge. Consequently, each of the chapters asks a question and then attempts to answer it. Another scholar with a different background and area of expertise would likely ask different questions or answer these ones differently. The point is not to provide the definitive answer, but to shake up the debate, disrupting our innovation studies status quo.

In Chapter 1, I ask why we should study innovation and what value an archaeological approach has. In this chapter, I emphasize the politicization of innovation, the value attached to the concept in contemporary contexts, and the way this valuation affects our ability to assess innovation in the past. I build on the long and quite fraught history of anthropology and archaeology in Tasmania in order to demonstrate that archaeological narratives of innovation are politically potent and socially constituted. Archaeologists, I argue, have a particular insight into the question of innovation because of our deep timeframe, but also because we are used to reconstructing worlds and logics where the common-sense solutions of the contemporary world do not apply.

Chapter 2 makes the case that, even given the limits of the archaeological record, innovation can only be understood through an explicitly social lens. In this chapter, I build outwards from a discussion of the changing narratives of early agriculture studies – from positivist to revolutionary to complex mosaics – in order to explore how archaeological model-building around innovation works. I contrast my social approach with the influential evolutionary school of thought and argue that the latter flattens a complex and difficult past.

The question of how innovations happen and whether archaeologists can understand invention is addressed in Chapter 3. I return to my first area of research, the invention of metallurgy in prehistoric Eurasia, to discuss where archaeologists can look for evidence of invention and how invention operates as a social phenomenon. A key point is that inventions, often conceptualized as ephemeral or momentary, exist in and emerge from complex networks of people, things and ideas; and these latter are discoverable.

The logical follow-on from a chapter about invention is one about adoption, so in Chapter 4 I ask why people might innovate or adopt innovations. To explore this question, I dig into the wide literature around the innovative practices of colonized people confronted by European hegemonic expansion. The point here is to build a story about innovation that moves us away from models of rational actors and savvy operators and towards a more socially contingent narrative of adoption.

In Chapter 5, I ask how innovations are communicated, both in terms of their transmission between generations or between experts and novices, and more broadly about their dissemination among communities. I use the example of the spread of Lapita practices and technologies in the prehistoric Pacific to compare narratives of migration and diffusion. I dig into the wider literature around apprenticeship and social learning and compare this with the more biologized narratives of knowledge transmission and meme theory.

Although this book is centered around innovation, most innovations fail or are rejected. In Chapter 6, I ask why people might resist or reject innovations and how conservatism operates. Although an absence of innovation is hard to pin-point in the archaeological record, I explore the example of later prehistoric Cornwall, where patterns of long-term continuity in social structure and settlement practice seem to allow locals to resist social, political, and economic changes associated with the Roman invasion of Britain. I build on this case study to explore the logics of resistance, persistence, and tradition in both living and ancient societies.

As a complement to the previous chapter's discussion of conservatism and resistance, Chapter 7 considers creativity and innovativeness. In this chapter, I explore the ongoing debates about the origins of creative behavior and creative problem solving during the Pleistocene in order to illuminate how archaeologists conceptualize creative and innovative practices and from where we see them emerging. I argue that, while creativity is a universal human trait – and likely one shared by a number of our hominin cousins as well as our own direct ancestors – the impetus to engage in innovative behavior is much more complex and socially contingent. I draw on multispecies ethnography and evolutionary archaeology to discuss how the world in which we live shapes our predisposition to innovation.

Chapter 8 concludes the volume by discussing the changing contemporary understanding of innovation and its social role. I summarize the arguments made throughout the rest of the book and look to the future of archaeological innovation studies with an eye to demonstrating how a social model of innovation offers both a means to critique our contemporary attitudes to innovation and technological change, and also a productive path forward for archaeologists interested in how and why people made the choices they did – at any point in the past.

1

Innovation as discourse

Why study innovation and why do so as an archaeologist?

Archaeology has a longstanding and probably inescapable fascination with the temporality of change. Historically, these narratives were often teleological: how did we get *here* from *there*? Early archaeologists took cues from geology and tried to understand the archaeological record through uniformitarian principles: that is that processes in the present could be used to explain patterns from the past (Trigger 2006, Chapter 4). In practice this meant, among other approaches searching for parallels in technology, social structure, belief system, and other practices between the ancient people whose things archaeologists uncovered and contemporary non-European populations encountered by European colonizers and settlers. From the late nineteenth century, evolutionary models dominated the emerging human sciences of anthropology and archaeology, a shared element of the otherwise quite divergent American and European archaeological spheres. Even as archaeological thought has fragmented over the last several decades – with new interpretative approaches emerging almost as fast as new scientific methods for understanding, identifying, and dating archaeological materials – how and why new ideas emerge and spread has remained a central concern of archaeologists around the world.[1]

Studies of innovation (past and present), of course, are not purely an archaeological endeavor. Focusing on innovation, as ambiguous and loaded as the term is, allows me to engage more obviously with the wider trans-disciplinary dialogue about invention, innovation, and technological and social change to which this book is both a reaction and a sort of answer. Archaeology, I would argue, has a special kind of insight into how and why people change their social practices, adopt new technologies, or refuse to do either. Ours is the only field that can investigate the whole life cycle of an innovation, from invention (if we are very lucky) through the various stages of adoption (and rejection) until it falls out of favor and/or is replaced by something else. Yet, at the same time, while archaeological discussions of technology are typically both incredibly detailed and highly precise, they are also frequently politically naive. In other words, although I believe that archaeological approaches to material culture and technology have great potential to enhance our understanding about the process of innovation and its place in human society, their relevance to studies of the contemporary world, to the population at large, and of course to non-archaeological research into innovation is rarely made clear.

In this chapter, I present the idea of innovation, and how it developed and continues to develop in scholarly and public discourse. Interpretations of innovation and innovative behavior drawn from the social sciences and framed by the short history of capitalism form the core of this discourse; and many archaeologists, whose material often derives from far earlier periods or regions largely unaffected (at least at the outset) by capitalist relations or industrial production, draw on it relatively uncritically. I start by examining the impact of applying a historic, Eurocentric model of innovation to non-Europeans, in this case Aboriginal people from lutruwita (Tasmania) and their society. This case study allows me to draw out the innately political core of our innovation discourse and suggest how a more critical archaeological perspective might allow for a deeper discussion of technological and social change in both the past and the present.

The strange case of Tasmanian technological change

In 1971, Rhys Jones submitted his Ph.D. to the Australian National University and, without meaning to, set the first sparks of a conflagration that would engulf the Australian archaeological community for years and that continues to flicker to this day. Jones spent nearly a decade in Tasmania running the first scientific archaeological excavation in that island's history within two caves on the north coast of the Bass Strait, within which he recorded stratigraphic sequences indicating sporadic occupation since c. 9,000 bp. Among his best-known results was the striking shift away from fish consumption between 3,500 and 3,900 years ago (Jones 1971, 1978). Although the bones of scaled fish were present in older layers, they seemed to disappear rather abruptly, a result he attributed to cultural changes among Indigenous Tasmanian communities rather than to changing environmental or economic factors. In fact, he found this shift away from fish consumption – which he had estimated as 20 percent of the pre-fish-abandonment diet – to be evidence for irrational and disadvantageous decision making.

To explain this archaeological pattern, Jones looked beyond diet and noted that layers without fish bones also lacked bone tools, which he believed were used for producing leather attire. Alongside longstanding observations on the "simplicity" of the Tasmanian lithic assemblage, he saw these economic, social, and technological changes as evidence for an ongoing disappearance of knowledge and skills. According to Jones (1977), Tasmanians did not simply shift their diet, they lost the technology to catch large fish, just as they had lost the technology to make bone tools and, as a result, to transform leather into clothing. In subsequent publications, Jones went on to argue, based in part on rather dubious ethnohistoric reports, that Tasmanians also forgot how to make fire. He framed these changes as "degeneration" and suggested that they contributed to an increasing "maladaption" among Tasmanian people. He went on to suggest that this maladaption set up the rapid disintegration of Tasmanian society in the face of colonialism, and that, even without European contact, Tasmanians were "doomed to a slow strangulation of the mind" (Jones 1977, 202). These observations were not limited to scholarly discourse, as Jones, the results of his Ph.D. research, and his specific interpretation of Tasmanian cultural "degeneration" were central to *The Last Tasmanian* (Haydon 1978), a film that was lauded by critics and awarded a Logie, even as Indigenous men and women fought against its portrayal of pre-colonial Tasmanian culture as degenerate, de-evolved, and extinct (Smith 2004, 180–2).

Jones' publications were not the first to argue that Tasmanian culture was "degenerate" or "de-evolved." In fact, his work added to a long history of European writing about the technology and society of Indigenous Tasmanian people. Colonialist, imperialist, and frankly racist discussions of Indigenous Australia feature in most of the foundational texts of anthropology and archaeology. In these works, written within the framework of European colonization and Anglo-centric racism, Indigenous Australians were portrayed as savages, their material culture and way of life was approximated to the European Palaeolithic, and their society was presumed to be largely unchanging and backward (McNiven and Russell 2005, Chapter 3). Even so, the treatment of Tasmanian people in these narratives was extreme: Tasmanians were believed by many – even in the face of contemporary skepticism – not to have known how to make fire (Gott 2002) nor to have had the technology for making clothing or the imagination to make art (Edward Tylor in Roth 1899, vi). In the social evolutionary models current at the time, Tasmanians were inevitably the lowest stage of human development, the least evolved, and the best evidence that non-European people were living fossils of ancient times – a position that was used effectively to absolve their British colonizers of their genocide (Davidson and Roberts 2008, 22–4). This depiction of Tasmanians, which was endorsed by Jones' research, positions them against the tide of progress: not just conservative but de-innovative, in the grip of both technological and social regression, and thus both less civilized and less human than even other Indigenous Australians. As we shall see, the tight linkage that developed during the Industrial Revolution between innovativeness – particularly technical inventions – and social valor continues to echo in our narratives of innovation and conservatism in the present day.

More recent archaeological, historical and anthropological research in Tasmania has called many of Jones' interpretations into question. As Hiscock (2007, Chapter 7) and others note, Jones vastly overstated the significance of scaled fish in the Tasmanian diet. He largely disregarded the evidence – in the form of thick midden layers – for intensive exploitation of coastal shellfish, a much more secure and easily accessed food source, especially since 3,900–3,500 years ago (Sim 1999). Closer readings of some of the historic journals and travelogues Jones used to write his thesis have led historians to question his interpretations of them, and suggest instead that fishing, among other activities, was carried out in the later Holocene, perhaps through the use of traps (Stockton 1982; Taylor 2007) (Fig. 1.1). Moreover, even if there were a cultural prohibition against eating fish, simply avoiding a major local protein source is not inherently irrational or maladaptive behavior – as Timothy Taylor points out, protestant British people themselves avoided fish for several centuries because of its Catholic connotations (Taylor 2010, 48), and recent isotopic research in prehistoric northern Europe suggests that a shift from maritime to terrestrial protein sources (that is, a shift from fish to meat) was a key part of the underlying cosmological changes linked to the adoption of agriculture at the start of the Neolithic (Schulting and Richards 2002). Challenges based on fuller readings of ethnohistoric sources and a greater abundance of archaeological data have also been made to Jones' claim that fire-making was unknown (Gott 2002; Taylor 2008), to the idea that a lack of bone awls necessarily indicated a lack of clothing (in fact, Bowdler and Lourandos [1982] suggested that bone tools are more common in phases with more fish bones, suggesting a use in fishing activities), and to the very idea that pre-contact Tasmania was lacking in innovation. As Taylor (2007, 11) notes, eighteenth-century Aboriginal Tasmanian culture, when not viewed through nineteenth-century European social evolutionary lenses, in fact appears to have been extremely dynamic, seeing the rapid adoption of new technologies (guns), materials (woven cloth, glass, ceramic, rust as a colorant), and practices (hunting with dogs, eating European foods and previously unexploited local foods, such as freshwater mussels), as well as the development of new songs and dances to contextualize them (Murray and Williamson 2003, 320).

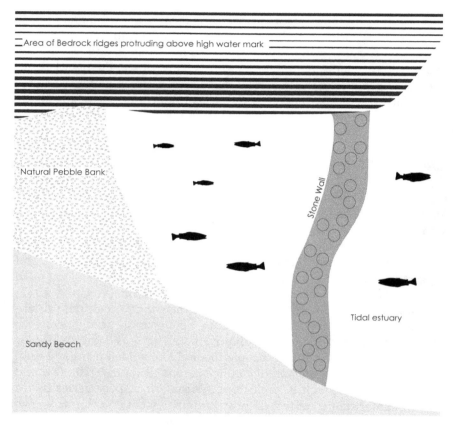

Area of Bedrock ridges protruding above high water mark

Natural Pebble Bank

Stone Wall

Tidal estuary

Sandy Beach

Figure 1.1 Schematic diagram of a stone fish trap in a tidal estuary built by Aboriginal Tasmanian people.

Yet the narrative of the culturally stagnant and technologically decrepit Tasmanian persists in the literature and in popular imagination. Certainly, *The Last Tasmanian* must take some of the blame, as it was among the first narratives of Indigenous Australia to be widely appreciated by White Australians; but, seeing as V. Gordon Childe's nearly century-old "Neolithic revolution" model of the early spread of farming also remains current in the popular imagination, and is still regularly taught in schools as current research (e.g., NCC 2013; see below and Chapter 2 for a wider discussion), perhaps we cannot expect there to be a rapid shift in public understanding of Tasmanian archaeology and history. However, what is more striking is that, despite repeated debunking by archaeologists and historians, the myth of Tasmanian regression continues to be cited in academic and specialist literature. Noted evolutionary archaeologist Joseph Henrich (2004) uses the case study of Tasmanian "loss of useful arts" to develop a model of "maladaptive loss" that he links to the demographic constraints of isolated communities. Notably, he does not weigh in on the various attempts to estimate pre-contact Tasmanian population size but relies heavily on Jones' interpretations without reference to more recent archaeological investigation. That this particular myth should continue to crop up even in the face of overwhelming contradictory evidence and its repudiation by contemporary Australian archaeologists probably lies in the ease of comprehending a simple narrative that also appeals to the confirmation bias of people for whom the belief in technological progress; the centrality of innovation to social and economic success; and, concomitantly, the perils of technological regression are an implicit part of their (post-industrial, western, capitalist) world view. The success of books such as Jared Diamond's *Collapse* (2005) and *Guns, Germs, and Steel* (1997a) certainly suggests that non-specialists prefer simple narratives that draw heavily on western ethnocentrism and discredited science, even while specialists bemoan them as "prescriptive cookbooks for cultural success that gloss over, and at times, disregard the details that social scientists have amassed over the past few centuries regarding the human condition" (Porter 2010, 149).

That there are popular narratives about ancient innovation that are well known and believed in spite of contradictory data makes clear that this is not just an academic debate: it has had profound implications and continues to have them in the present day. In Tasmania, as Smith (2004, 181–2) explains, the popular narrative of Tasmanian degeneration and extinction fed directly into political policy debates, undermining Indigenous Tasmanians' ongoing claims for land rights and legal recognition, and culminating in the Aboriginal Relics Act 1975, which declared that only material made prior to 1876 would be considered a product of Aboriginal culture. Similarly, Wilcox (2010) argues that Diamond's simple, environmentally deterministic narratives of western technological dominance and conquest are "potent instruments" of oppression that use the seemingly objective language of science to justify, de-personalize and de-politicize European colonialism and colonialist interpretation of Indigenous sites and material culture.

This linkage between perceived technological capacity and the value of a given society is, of course, not a recent one. Perceptions of technological superiority, of greater capacity for innovativeness in particular, play a crucial role in the European narrative of justified colonial and imperial expansion (Adas 1989). Both Jones' Tasmanian research (in print and on film) and Diamond's widely read (and Pulitzer Prize-winning!)

accounts of European expansion and social collapse echo this long, ugly history by centering their arguments around the innovativeness of Indigenous people and specific European inventions, implicitly tying together perceived technological creativity and innovation with ideas about social structure, worthiness, and cultural agency. Such views, unfortunately, extend to almost all Indigenous populations, with similar deleterious results as in the case of Indigenous Tasmanians. Colonialism and colonial narratives of power, dominance, and racial difference are deeply rooted in European culture and politics – early modern readings of Roman history were used to justify the foundation of English colonies in sixteenth-century Northern Ireland (Horning 2015). These ideas continue to resonate in the present, not just in the relations between colonized people and settlers, but also in the very foundational structures of business and accounting practices that bolster modern innovative industries and that, themselves, originated in the account books and management structures of slave plantations (Rosenthal 2018).

The idea of innovation, then, is not value-free, and archaeological narratives of innovation are not necessarily objective. Thus, before we can begin to deconstruct the act of innovating, let alone the presence or absence of innovative practices and technologies within various societies, we must first develop a better sense of how to conceptualize innovation and its place both within society and within academic dialogues about technological change and creativity.

Identifying and examining innovation in the present

Innovation is a common topic in the present day. We live in a world where innovation, innovativeness, creativity, and invention are all highly positive personal and societal traits. They are buzzwords, almost laughably over-used by everyone from corporations to universities to government-granting agencies to non-governmental organizations, charities, and politicians. The idea of innovation that we use today – one that centers on technological novelties, associated practices and thought processes – is a product of the modern era. As Girard (1990) points out, prior to 1700, *innovation* was typically defined as dangerously unorthodox religious thinking or practice that challenged the fundamental principles on which the world was based. This conceptualization of the act of innovation as a subversive threat to the status quo has been traced back to Antiquity, though the earliest uses of the Latin *innovo* (renew, rejuvenate) were more ambiguous (Godin 2015b). Only with the Enlightenment shift away from theistic conceptions of the universe, of social structures, and of political systems, and towards a more rationalist, scientifically informed idea of the world and humanity's place in it did innovation begin to take on its modern usage. Over the course of the eighteenth and nineteenth centuries, the invention of new technologies and technological creativity itself, particularly within the context of capitalist industrial expansion, became inextricably linked to moral and social progress. Joseph Schumpeter's (1943) economic writing made innovators and entrepreneurs into Culture Heroes; and the linkages between innovation and the advancement of society were only tightened and reinforced during the so-called "golden age of capitalism" in the second half of the twentieth century. In 2019, Google returns 12 million results for the search string "Culture of innovation," every single one of which (on the first ten search pages, at least) uncritically positions a culture of innovation as a positive thing – and one that should be nourished and promoted when present, and bemoaned and sought after when absent. Today, "innovation" is a solution to global problems, such as climate change (to which, of course, it contributed and continues to contribute),[2] as a factor mitigating the potentially damaging aspects of globalization (Archibugi and Iammarino 1999), and as a solution to educational quandaries (Schlechty 2001). It almost does not matter what the problem is, whether political, personal, or economic: we are told more innovation can solve it (e.g., Feenberg 2002, 155–8).

This seeming constantly advancing march of technological progress and ever-increasing innovation, this normalization of technological revolution, is of course the product of a concerted effort by generations of industrialists, corporations, politicians, and policy makers rather than the natural state of the world. As Robins and Webster (1999) remind us, early attempts to resist the holistic alteration of the social fabric in the guise of technological innovation by industrial capitalists were punished swiftly and viciously. The infamous Luddites, nineteenth-century British agitators who, through radical action and destruction of machinery, protested the encroachment of the Industrial Revolution on traditional ways of life, are often caricatured, even today, over 200 years later, as mouth-breathing technophobes who attacked what they did not understand. Their uprising was brutally suppressed. The Destruction of Stocking Frames, etc. Act 1812 was passed in response to the Luddite uprising, making destruction of mechanized looms a capital crime. The British Army was called up to corral Luddite protests, and elaborate show trials resulted in the execution and transportation of dozens of suspected Luddite activists. This disproportionate response to protest and property damage is a reflection of the times – there was not yet a union movement in the 1810s and any protest by the working classes was seen as a major threat to the established order – but, in Robins and Webster's (1999, 49) words, more tellingly it also "exposes and challenges the modern mythology of progress." If the onward push of innovative progress were really as robust and inevitable as it is currently perceived, the Luddite uprising would not have needed such violent and total repression; it would have fizzled out on its own without hope of success. Thus, the popular view of innovation and its place in society is colored by a sort of techno-fetishism or "technoromanticism" (*sensu* Coyne 1999) that is as historically situated in our networked and global-ized world as the Luddite concerns about mechanization, urbanization, de-skilling, and loss of traditions were in the time of Enclosure and early industry.

The view of innovation from within the academy is somewhat more nuanced. Innovation research developed rapidly in the social sciences and human sciences after World War II. The vast majority of this writing drew on Schumpeter's foundational models of capitalist innovation and took a distinctly evolutionary perspective, suggesting that those inven-tions that are "most fit" for solving a given problem or problems are also most likely to be widely adopted as innovations. This research was carried out in dialogue with policy makers and developed out of an intellectual framework that viewed technological changes (and the concomitant changes in social practices that accompanied them) as an unmitigated good (Godin 2010a, 2010b), a perspective that remains strongly present in contemporary research in innovation studies, such that one leading voice in the field can suggest without irony that "there seems to be something inherently 'human' about the tendency to think about new and better ways of doing things and to try them out in practice" (Fagerberg 2006, 1). In effect, this research, which today is largely carried out by economists and researchers in organizational and management studies, builds models of innovation designed to facili-tate modernization of technical processes; promote the uptake of new inventions; increase commercialization of new ideas and products; and, thus, stimulate economic growth that is positioned, at least implicitly, as improving quality of life for the general population.

At the same time as the economic strand of innovation studies was developing, a human-sciences approach to technological change and the uptake of new ideas and practices also emerged. Foundational research in anthropology (Barnett 1953), sociology (Rogers 1962), and cultural geography (Hägerstrand 1966) examined the human and cultural side of invention and the spread and uptake of innovations. While this research was also highly influenced by positivist attitudes to science and techno-logical change – Everett Rogers (1962, 1) starts his book by explaining that, in order to speed up the diffusion and adoption of innovations, one must understand the process – it is distinguished by an attempt to understand innovation from the bottom up, and to contextualize the adoption of new practices and technologies. Barnett, Rogers and those they influenced highlighted the role that localized value systems play in the adoption or rejection of innovations; and, perhaps more significantly, they made clear that adoption is but one potential outcome (and by no means either the most common or the most rational) of the extended innovation process (see Rogers and Shoemaker 1971).

In contrast to the de-personalized and pseudo-evolutionary models of the dissemination and adoption of innovations promulgated within the social sciences, the work carried out in the human sciences emphasizes that trajectories of innovation adoption are not linear, but rely on a complex constellation of distinctly social factors, including the control and dissemination of knowledge at the local level, the presence of particularly influential members of the community, and the ability of potential adopters to come to grips with a given innovation through a slow process of testing and experimentation. Although widely cited, Barnett's work does not seem to have led to a flourishing of anthropological studies of innovation, nor do many archaeologists – in a discipline more traditionally concerned with technological change – seem to have followed his lead in studying the social aspects of invention and innovation (see below). By contrast, Rogers' pioneering diffusion research remains prominent in academic and professional contexts; *Diffusion of Innovations* is currently in its fifth edition (Rogers 2003), and his conceptualization of the innovation process and the stages of adoption have been adopted into management and organizational studies as a toolkit to speed innovation uptake (Moore 1991). Nevertheless, what is clear is that the idea of innovation itself – as well as the actual process of creating and adopting or rejecting a new thing, idea, or practice – does not exist outside specific social contexts and is crucially linked to relationships between individual people.

Yet, despite their differences, both of the two major strands of innovation research, the economic and the sociological, share a similar conceptualization of innovation itself, which is almost universally synonymous with technological change and linked to increased economic growth. Although the more socially embedded approaches advocated by sociologists and geographers undermine the widespread image of technological change as a neutral and natural process, they still largely retain a strongly positivist outlook framed around a distinctly western, industrial capitalist framework in which innovation necessarily implies both successfully adopted technological changes and economic growth (Blake and Hanson 2005, 682). Within this framework, it is standard to separate out different sorts of innovations based on their character and perceived impact; so we find ourselves talking about *product innovations* (i.e., the creation of new products) and *process innovations* (i.e., the creation of new organizational structures, largely within firms) that can be either *incremental innovations* (small changes that do not disrupt the larger economic and technological systems) or *radical innovations* (large changes that create whole new categories of product, technology, skill set, etc.) (OECD and Statistical Office of European Communities 2005). Whilst these seem like quite general terms, in practice it is clear that, when social scientists use "innovation," they do so primarily as a shorthand for "technological innovation," by which they mean "manufacturing" (Blake and Hanson 2005, 681).

This conflation of terminology has been examined in considerable detail in feminist studies of innovation and entrepreneurship, which make clear that innovation is a highly masculinized concept, and that an aura of innovativeness tends to adhere more to traditionally male sectors – IT, manufacturing – than to female-dominated areas, such as healthcare or the service industry (e.g., Alsos *et al.* 2013; Blake and Hanson 2005; Crowden 2003; Wajcman 2000). Essentially, this research highlights two fundamental problems with the innovation studies canon and its use in policy making: first, to a much greater extent than is typically acknowledged, our perception of a new thing or idea as innovative depends on the social identity of the innovator and the local context in which their innovations are adopted; and, second, innovations other than successfully marketed changes to technology and manufacturing are given short shrift, if they are discussed at all.

The focus on influential individuals within the innovative process – for Schumpeterians, the entrepreneur, for Rogersians, the early adopter – is consistent with the narratives of heroic inventors that form a key social element of the Industrial Revolution. From James Watt to *Wallace and Gromit*, stories of the solitary and highly dynamic inventor permeate popular culture and mainstream histories. In these stories, the inventor (or the entrepreneur in the twentieth-century versions) is a Great Man (unsurprisingly, they are almost always male), identifying problems and developing creative technological solutions that are immediately effective and catalyze further change and, often, major industrial developments. Thus, Watt's steam engine holds greater narrative weight in histories of innovation than contemporary process innovations in consumer goods manufacturing and food production, such as the first implementation of production lines or the development of new processes for timing operations and for developing and maintaining efficient national distribution networks (Berg 2010; Bruland and Mowery 2006).

What ties these less celebrated sectors together is that they are, by comparison with textile manufacture, low prestige – being marketed to farmers rather than industrialists, and to women and their households rather than men and their companies – an attitude that is not confined to the nineteenth century, since studies of innovation remain largely concerned with "male heroes, big projects and important organisations" (Wajcman 2000, 453). Innovation research regularly (although not universally) positions male-dominated industries and fields as particularly innovative, while assuming women and other minorities are passive consumers of new technologies. Innovations in feminine fields such as service industries or the public sector, and in small businesses and firms focused on local rather than international markets, are rarely recognized as such (Alsos *et al.* 2013; Blake and Hanson 2005). In fact, even when women are present within industries, firms, or fields recognized as innovative, their innovations may be actively resisted within masculine spaces. Crowden's (2003, 26–8) focus groups of women in biotech and multimedia corporations reported that their credibility was questioned by male colleagues, their contributions to brainstorming activities were rejected or resisted by male-dominated workgroups, and even incremental process innovations they developed were only rarely implemented. Similarly, in their recent book on the history of computing in Britain, Mar Hicks (2017) observed that, because the technical workforce that supported the mid-twentieth-century computing industry was feminized, skilled labor was not recognized as such, pay and working conditions depreciated, and the industry stagnated as a result.

In innovation research, a focus on seemingly isolated or otherwise extraordinary, frequently cis male, individuals as key vectors for innovation is not limited to technological contexts or industries. For example, although well-known (male) authors such as Shakespeare or the sixteenth-century French poets known as "La Pléiade" are generally treated as important sources of linguistic innovation, socio-linguists have demonstrated repeatedly that most linguistic innovations – new words, new inflections, new styles of vocalizing – are developed first by adolescent women in their conversations, spoken and written, with each other (Cameron 2003; Labov 1990; Tagliamonte and D'Arcy 2009). Of note is that as long as these developments are largely confined to female speech they tend to be resisted and even actively derided by males and elites, but once they begin to be adopted into male speech patterns (usually with a delay of about one generation) they become neutral or positive developments (Tagliamonte and D'Arcy 2009, 63–4).

Since Great Man narratives remain central to innovation studies, and technological innovation itself is often considered to be an unalloyed good to be promoted, it is unsurprising that research on radical product innovations – the invention and adoption of new technologies that appear to cause a great leap forwards in associated technological systems – is much more common than studies of incremental or process innovations, even when these latter can be shown to have had a greater long-term impact (Kline and Rosenberg 1986). Of course, process and incremental innovations are also forms of innovation, which, as we have seen above, are often associated with lower-prestige industries or innovators. Feminist analyses of innovation in organizational contexts suggest that innovation studies should be oriented away from this traditional focus on the invention and impact of emblematic new technologies, and towards the incremental and process innovations that contextualize their development and spread (e.g., Alsos *et al.* 2013). These researchers advocate that we shift our focus away from industrial-scale manufacturing and towards the efficient and innovative delivery of services, or the localized economic benefits of small-business women meeting local demands with insider knowledge in ways that large corporations are structurally incapable of delivering (Blake and Hanson 2005).

Yet, I would suggest that re-orienting innovation studies away from manufacturing and towards social or process innovations still misses a key limitation on the field and its ability to draw out and develop a broader understanding of the innovative process: namely, that the study of innovations, like the idea itself, has developed almost entirely within the framework of capitalist relations of technological development, production, and consumption in which all innovation, especially technological innovation, is explicitly conceptualized as leading to economic growth and, thus, to more generalized social benefits. Although feminist perspectives rooted in Marxist conceptions of production and consumption have delivered an effective critique of mainstream innovation studies by broadening our understanding of technological change and its social position (Wajcman 2000, 2009), this research remains anchored in the present and, consequently, is largely responsive to these capitalist framings of innovation rather than opening a wholly separate approach.

It is my position that, to develop truly holistic and comprehensive understandings of innovation – of why and how people innovate, of how they talk about innovations and why they resist or reject them or, for that matter, why one group of people seemingly gleefully adopts innovation after innovation while their neighbors do not – we must look beyond traditional fora of innovation and towards the pre-capitalist past. Recent work in ancient history has emphasized the deep roots of modern concepts of innovation (Godin and Lucier 2012), and there is no question that significant social and technological changes characterize the pre-modern world (though perhaps at a smaller scale and slower pace than in our globalized, digitally networked present). Archaeology, then, with its deep-time perspective and variety of intellectual and methodological approaches for disentangling the people, practices, and technologies of humanity's past, seems the obvious field from which to develop this research. Moreover, archaeologists, along with other historians of

technology, have worked hard to develop interpretative tools that identify where and how present-day understandings of material culture,
technology, and social structure influence our picture of the past –
particularly, the past for which we have no written records – so that we
might do more with our research than simply reconstruct the present in
the past (Binford 1992; Shanks and Tilley 1987). Nevertheless, archaeological approaches to innovation necessarily retain technological change
as a central part of the innovative process, as these tangible traces
of human action and relationships are often all that is left for us to
study. Yet, this same limitation applies to most archaeological research
and has not prevented us from developing an understanding of complex patterns of social and political relationships, cosmology, or wider
technological systems. Moreover, the interdisciplinary technological and
material-cultural "turn" of the last several decades (Hicks 2010) has
made clear that social relations, objects, technological systems, and the
environment in which they all develop cannot be so easily disentangled
and, thus, that each may give insight into the others.

Technology, technological change, and human society

Traditional and mainstream discussions of technology tend to treat it as
occupying a position separate from human society. People manipulate
things and do so in ways that can have profound effects on other humans,
on social structures, and on future generations, but the things themselves
are passive, inanimate objects. Moreover, we tend to assume that the
functionality of those objects – their capacity to fulfill the physical or
other uses to which they are put and for which they were designed – is
their primary and most valuable attribute. Thus, technological change
happens when a new tool is needed, or a more efficient or better functioning one is invented. Moreover, the corollary is that technological
change is directional, in that it develops in a linear fashion, with each
new technology or tool improving on the previous one. This model
of linear technological progress under-pins, and developed alongside,
evolutionary models of human social development that were originally
fleshed out in the nineteenth century and continue to resonate today.

In their original forms, evolutionary frameworks suggest that specific
levels of technological development and, in some cases, specific types of
tools or technological systems can be linked directly to different social
formulations with increasingly sophisticated technologies being indica-
tive of increasingly sophisticated social structures. Thus, Sven Nilsson's
1868 treatise *The primitive inhabitants of Scandinavia* (Nilsson 1868)
characteristically conflated social organization and technological attain-
ments to suggest that human social progress could be charted from

Savages to Barbarians to Agriculturalists to Civilized. In the following year, Edward Tylor, known now as the father of British social anthropology, published his celebrated monograph *Researches into the Early History of Mankind* (1865), in which he laid out a similar model of social development linked explicitly to technological progress. According to Tylor, Savagery could be characterized by a hunter-gatherer lifestyle, Barbarism by the early stages of agriculture and animal domestication, and Civilization by the development of writing. Tylor elaborated this model by comparing the various technological systems of living indigenous peoples to archaeological assemblages collected in Europe, and argued extremely influentially that ancient European people could be compared directly to modern societies living in the states he described as Savagery and Barbarism. American anthropologist Lewis Henry Morgan (1985 [1877]) subsequently elaborated on this model, distinguishing phases of lower, middle, and upper Savagery, Barbarism, etc., each associated with specific technological innovations and social formations in an evolutionary sequence (Table 1.1).

The conception of linear progress encapsulated in these early evolutionary models – that is, that successful innovations are more fit than those they replace and that technological change is a product of conscious experimentation aimed at improving functionality – is still the guiding principle behind most models of innovation (see Pfaffenberger 1992). Although rigid, unilinear evolutionary modeling ceased to dominate the human sciences in the early twentieth century, its impact, albeit largely shorn of the obvious racist, ethnocentric, and colonialist heritage of nineteenth-century research, can still be seen in influential work by White (1959) and Sahlins and Service (1960). Yet, even as the interpretation of social change has become more nuanced, it is only in the last several decades that the many assumptions about technology – and in particular about its relation to human society – tied up in these nineteenth-century models have begun to be questioned.

The interdisciplinary field of technology studies began to coalesce in the 1970s and gained momentum throughout the 1980s and 1990s. The idea of innovation was central to this rapidly shifting intellectual environment as researchers looking at the trajectory of technological changes in the nineteenth and twentieth centuries began asking "Why did this thing catch on, rather than that equally useful one?," and "Why did it take so many years for a process to develop that seems logical in hindsight?" Although researchers allied to this emerging area of study came from a variety of backgrounds, a few shared central concepts rapidly materialized that tied together the sociological studies of science with the anthropological and historical studies of technology in human society and feminist narratives of women's participation in industrial

Table 1.1 Tylor and Morgan's evolutionary models of social and technological development set against the standard European archaeological periodization, which was developed contemporaneously.

Tylor	Morgan		Archaeological ages
Evolutionary stages	Evolutionary stages	Level of technological development	
	Lower Savagery	Fruits and nuts, speech	Palaeolithic
Savagery	Middle Savagery	Fishing and gathering, use of fire	
	Upper Savagery	Hunting and gathering, bow and arrow, spear	
	Lower Barbarism	Horticulture, pottery	
Barbarism	Middle Barbarism	Animal domestication (Eurasia), maize cultivation (Americas), irrigation, bronze smithing	Neolithic
	Upper Barbarism	Cereal cultivation, iron smelting, wheeled vehicles, potter's wheel, loom weaving, poetry	Bronze Age
	Ancient Civilization	Iron-pointed plows, animal traction, coinage, hieroglyphic and phonetic alphabet, writing, cities	Iron Age/ medieval era
Civilisation	Modern Civilization	Telegraph, power loom, steam engines, printing, gunpowder, photography, science democracy	Post-medieval era

workplaces. Primary among these is the idea that technology and things do not exist outside the sphere of human relations but are, in fact, enmeshed with culture (Feenberg 1995). What this means in practice is that material culture cannot be treated as separate from the processes through which it was produced, and that these processes, in turn, emerge from the wholly social world of human relations. Following the anthropologist Bryan Pfaffenberger (1988, 241):

Technology can indeed be defined as a set of operationally replicable social behaviours: no technology can be said to exist unless the people who use it can use it over and over again. To the extent that technological behaviours are replicable, the interpenetration of physical elements (e.g., tools, resources, etc.) and social communication (diffusion, apprenticeship, etc.) is presupposed.

In other words, *technology* as a concept comprises material culture, techniques, production processes, operational knowledge, and human social values – a heterogeneous complex of actors, motivations, and processes, which Hughes (1987) refers to as a *socio-technical system* (for a similar model built from francophone technology studies, see Lemonnier 1986, 1992).

The social element of technological systems means that technological change cannot be a product of evolutionary development or the steady improvement of functionality, but instead must reflect human choices, values, and the wider social context in which it occurs. Bijker (1987, 1995) frequently refers to the case of Bakelite, an early plastic invented in 1907 that became widely popular for the production of household items, jewelry, and non-conducting parts in small electronic devices from the 1920 to the 1040s. Prior to Bijker's research, the widely held view of Bakelite was that it was the inspired creation of the visionary inventor and chemist Leo Baekeland, as well as the first link in a chain of inventions leading inexorably to the proliferation of plastics that dominated late-twentieth-century material culture. However, through analyzing the social and technological contexts in which Bakelite was invented, produced, and adopted by the public at large, Bijker was able to demonstrate that Baekeland's invention and its successful adoption resulted directly from his participation in multiple socio-technical systems, giving him the necessary background in the production of synthetic resins to build on previous research, as well as an interest in celluloid film production that pushed him to experiment more widely with the properties of his own inventions and, crucially, experience as a businessman to allow him to move his product out of the lab and into people's homes. Moreover, not only was Bakelite not the immediate and obvious success that most post hoc stories of plastic invention would have it, but it took a decade of development plus the historic accident of sudden quantities of reduced-price phenol being made available after the end of World War I, as well as the emergence of consumer electronics (and, thus, the unexpected high value placed on a non-conductive, moldable resin), for Bakelite to begin to diffuse. As we shall see, this decidedly non-linear and deeply historically and socially contingent process is a far more normal part of innovation sequences than the linear process of research and design, leading to new ideas, leading to innovation and adoption, which remains enshrined in much of the innovation-studies literature.

Subsequent research along these lines has suggested not only that objects and technologies are socially constructed (see Pinch and Bijker 1987), but that the socio-technical systems themselves have their own "momentum" (Hughes 1994), which can affect the various groups of people bound up in them. So, for example, the expansion of the electrical grid in the USA was initially guided by political and economic decisions, but as its socio-technical system increased in size and complexity, the presence of electrical infrastructure began then to influence subsequent political and economic decisions about where future infrastructural developments should be constructed, as well as demographic patterns, leading to the clustering of new population centers in areas with pre-existing electrical grid infrastructure.

Social theorists have proposed that humans and objects exist in a network and that, within this network, objects have an ability to act on or mediate human relationships, just as humans can act on objects – a principle known as symmetry (Latour 1992, 1993; Law 1987). In this perspective, "human beings are not sovereign with respect to technology but are, rather, inextricably interwoven with it" (Verbeek 2005, 44). What this means for us is that, in studying technology, we are implicitly studying human activity and human choice (Lemonnier 1993) and, moreover, that changing technologies must be seen in light of the social networks in which they operate and that shape and are shaped by them.

In recent years, feminist and other scholars have pushed back on this framework, arguing that the principle of symmetry depoliticizes the complex relationships among people, things, and systems and removes the possibility of delineating experiences of oppression or marginalization (Davis 2020, Chapter 3 with references). Indeed, proponents of the approach recognize this failing, with Edwin Sayes (2014) recently arguing that politics and morality are fundamentally outside the scope envisioned for this interpretative method despite their salience to understanding a not insubstantial portion of the population's lived experience. Ernst Schraube's (2009) notion of technology as materialized action offers an important alternative to – or perhaps adaptation of – this approach, in that it accepts that people and things are interconnected and recursively constituted, but deems this relationship asymmetrical because agency rests only with humans. Objects and technologies shape the world around them, but people shape objects and technologies, infusing them with meaning, power, and direction. This approach allows us the interpretative space to observe and understand other asymmetries – of power, of gender, of privilege, etc.

More radical, perhaps, is the parallel observation that we ourselves are not entirely contained in our bodies, but also exist across networks of social relationships in which material culture plays a key role. Feminist scholars have suggested that the information age has seen an assault on embodied identity and the concomitant emergence of cyborgs or post-humans – people whose selfhood and cognition are distributed in networks that include human and non-human entities (Haraway 1985; Hayles 1999). Thus, shifting socio-technical systems implicitly also shifts our conceptions of ourselves, changing how we are constituted and how we forge relationships with each other. The recent flurry of research and popular inquiry around how increasingly popular social media technologies are (or are not) changing the ways we form and maintain relationships (e.g., Dunbar 2012; Wellman 2011) is a crystal-clear example of how, even within a very short time span, we are able to observe that changing technologies equally change how we relate to each other and how we form our public and private identities.

Yet, despite breaking down the intellectual systems that traditionally separated person from object and technology from society, the majority of this research retains an implicit understanding that complex human–technology relations are the product of post-industrial developments, and that a real division exists in how people and technologies intersect in the modern and pre-modern eras (or industrial and non-industrial contexts). Anthropologists would argue that this distribution of the self is not a product of post-modernity or late-capitalist power dynamics, but is rather an alternative and widespread form of personhood, though one unrecognized by western philosophical and legal frameworks. Marilyn Strathern (1988) famously argued that Melanesian personhood is dividual – that is, not bounded by the body but existing across a network – and partible, in that parts of one's personhood can and frequently are materialized and separated from oneself in the form of material culture and gift giving. So, in passing on heirlooms, favorite books, or family recipes we materialize an element of ourselves and allow it (us) to be incorporated into the selfhood of the person to whom we have offered it. Innovation, then, is an embodied experience with the potential to alter one's own sense of self and way of relating to other people, as well as the wider social and material worlds out of which identity and personhood are formed.

Some have even taken this idea one step further, adding time depth and allowing us to suggest not just that it is present societies where we find these networked identities but that humans and objects have always been entangled and, perhaps, that this very entanglement is what distinguishes our species. As Taylor (2010, 8), somewhat poetically, suggests, "we have never been wholly natural creatures, and we have evolved to be increasingly artificial." That is, in manipulating the environment and raw materials of nature through the creation of increasingly sophisticated tools, our ancient hominin ancestors unintentionally began the process of shaping our development as a species, freeing us from some of the constraints of evolutionary pressure and allowing a hairless primate, lacking adequate teeth or claws, to survive and, eventually, develop the large brains capable of complex, symbolic thought and communication that have characterized our genus for the last several hundred thousand years.

Archaeological readings of technological change and innovation

Archaeologists, by necessity, are experts in ancient technology and, more than that, experts in using ancient technology to tell the stories of past people and societies. Often, technology is the only clue we have about who people were, how they lived their lives, and what they valued. Consequently, over the last 200 years of sustained archaeological practice, a number of tools have been developed to make sense of the residue of past people, their material culture, and their socio-technical systems. For the vast majority of this time, technology and material culture were treated as external to social developments.

One of the great interpretative advances of the nineteenth century was the insight that, by comparing sealed contexts – that is, those that have definitely not been opened since Antiquity – chronological changes in material culture could be identified and the sequences in which new tool types, new forms of decoration, and new technologies developed could be sketched out. The development of these typo-chronologies gave order to the highly fragmented and variable archaeological record, and allowed early archaeologists to demonstrate, first, that meaning could be found from ancient materials and, second, that there was a long history of human activity that pre-dated written history (and, in particular, the Bible) (Schnapp 1996). Both of these were, at the time, controversial points. Within the larger intellectual climate of the nineteenth century, it was a logical progression to move from chronological changes in archaeological materials to evolutionary explanations; European archaeological materials were frequently used as comparanda to demonstrate that non-western people were so-called "living fossils," existing at the same level of development as Europeans had reached deep in their past (McNiven 2020). Thus, stone-tool-using Indigenous Tasmanians who used neither complex tools nor practiced familiar technologies, such as agriculture, were quickly identified by European invaders as savages, unchanging relics of a past world – a categorization that, of course, made their subordination to European colonists seem both inevitable and less morally objectionable.

Towards the end of the nineteenth century, a German school of archaeology developed the idea that different assemblages of materials indicated different groups of people, and that, in observing the changing geographic extent of these assemblages over time, we would be able to track the movement or migration of past people (Trigger 2006, 232–60). Vere Gordon Childe, famous archaeologist and public intellectual, used this concept to build a detailed and meticulous history of prehistoric Eurasian people, their movements, and their efflorescences, which made clear that archaeologists could do more than just make sequences of objects: we could start to tell the stories of identifiable groups in the past (Childe 1925, 1929). Childe was also an ardent Marxist, and he deeply believed that the invention of key technologies (i.e., agriculture, literacy, metalworking) would have shaken the established social and political foundations of ancient peoples' worlds, leading to revolutionary change and progress (Irving 2020). In this culture–history framework, society and technology are static unless an external pressure – a groundbreaking new technology, an environmental catastrophe, a large group of armed migrants – forces changes to occur. Where similar innovations appear in disparate areas, only migrations or invasions were thought to explain it. All innovation was seen as coming from outside, usually with new people whose way of life, material culture, and language replaced (often, implicitly, violently) pre-existing culture groups. This view of ancient cultures as fundamentally conservative and unchanging was a direct descendant of nineteenth-century ideas about non-industrialized societies, which suggested that invention was infrequent among nonwestern and pre-modern people. Thus, revolutionary inventions were believed to have been invented once and then to have diffused outwards as innovations from their point of conception, triggering major cultural changes as they were slowly adopted. This model suggested that, for example, agriculture was invented for the first and only time in Eurasia in the "fertile crescent" of Mesopotamia, and to have diffused outwards in all directions, bringing transformative change along with it in what Childe memorably named the "Neolithic Revolution."

However, by the later twentieth century, new scientific techniques were calling the old chronologies into question, making clear that many more fora of invention for key ancient technologies (including agriculture and metallurgy) existed, and opening up more tools for materials analysis. At the same time, a social evolutionary bent in anthropology called into question the rather static culture–history models pioneered by Childe and his contemporaries. Attention turned to systems theories of human society, in which social elements were nodes in systems alongside technology and material culture, both governed overall by environmental pressures (Clarke 1968). This framework allowed studies of ancient technology – that is, the processes by which people manufactured tools and the uses to which they put them – to emerge as it became clear that what people made; how they made things; and how the way they used the things they made reflected cultural, economic, political, and environmental concerns. In other words, single events such as migration or invasion were no longer understood to be robust explanatory models for the complex process of change and development testified to in the archaeological record.

That said, despite the increasingly central role that in-depth technology studies played in scientific archaeological research, technological change and innovation continued to be understood in a purely functionalist or evolutionary framework. Innovations were adopted when they served specific physical or economic functions or allowed a given group of people to thrive more efficiently in their environment. This highly scientific school of archaeological practice was subjected to a major critique in the 1980s based, in part, around the assertion that material culture studies gives us as much or more insight into symbolic and ritualized social values as into economic or adaptive behavior (Hodder 1982). These two strands of archaeological research, the scientific and the post-modern, are typically treated as constantly in opposition (e.g., in Johnson 2010) (although for an ultimately unsuccessful stab at synthesis, see Cullen 1995). However, the 2010s have seen them brought together in interdisciplinary approaches to material culture studies that seek to integrate the results of scientific analyses with social theories in a relational approach inspired by the literature on human–object networks and distributed personhood noted above (Harris and Cipolla 2017).

In fact, with the present material turn in the human sciences, it is clear that the in-depth technology studies, the symbolic and social interpretations of material style, and the systems models that linked people to things and to their environments in dynamic networks actually provide archaeologists with extraordinary interpretative and methodological tools to study human society, socio-technical systems, and changes in the two over time. Thus, for example, I and other colleagues have used close analysis of coeval stone and metal tools to demonstrate that the transition from the Neolithic to the Bronze Age – and concomitantly from a socio-technical system in which stone tools were of crucial importance to one in which they were largely replaced by metal – was neither a linear process nor an obvious one, but took place slowly in a punctuated fashion that differed from one cultural context to another (e.g., Bray 2012; Roberts and Frieman 2012; see Chapter 3).

Despite the necessary and close focus on technology, material culture, and technological change, within archaeology innovation has proven to be a particular problem. Although the discipline's very foundations lie in the understanding and identification of technological change (e.g., Rowley-Conwy 2007), it is only in the last few decades, largely in response (both positive and negative) to post-modern critiques of science and the scientific process, that archaeologists have begun to ask how and why innovations develop and are disseminated (e.g., O'Brien and Shennan 2010; Schiffer 2011; Van der Leeuw and Torrence 1989; Walsh *et al.* 2019b). While archaeologists were among the first to suggest that material culture and technology were socially constituted (Killick 2004; Lechtman and Merill 1977), and the social construction of technological systems has been widely discussed in archaeology over the last two decades (Dobres 2000; Gosden 2005), the literature on innovation is rather more limited. It rarely moves beyond a laser-like focus on origins: the identification of the first or oldest example of a given technology or tool type and its chronology of adoption. Thus, we see endless papers loudly trumpeting increasingly early dates for permanent settlements, cereal cultivation, or the domestication of dogs, but very few (and usually quite speculative ones) examining the social context of different areas with early agriculture or why and how agriculture was rejected in some parts of the world (see Chapter 2). Instead, innovations are still usually discussed in functionalist terms as fulfilling obvious economic or physical needs, with little thought given to why or how some new ideas or technologies become more widely accepted than others, and what processes allow for their adoption or increase the likelihood of their rejection.

Indeed, a considerable part of the current archaeological literature on innovation avoids all questions of social context or the interpenetration of people and things by positioning innovation as the mechanistic transfer of cultural data – an evolutionary process governed by fitness and demographic pressures. A fuller critique of this approach is presented in later chapters, but it is worth noting here that this evolution-inspired research often fails to distinguish between internal social logics promoting the adoption of an innovation and post hoc external assessments of functional or evolutionary improvement (Taylor 2010, 141). Although some archaeologists have proposed evolutionary models with more awareness of, and space for, social factors (Cullen 1993b, 1995, 1996a, 1996b; Jordan 2015), these are not the narratives that dominate in that field.

A central argument of this book is that innovation is inherently social and that the process of innovation – from the recognition of a new invention as desirable, through phases of experimentation and alteration to fit pre-existing cultural norms, to the ultimate decision to adopt or reject – can only be understood by looking at shifts in the wider sociotechnical system throughout this process. Moreover, *innovation* is itself a concept with a great deal of baggage, being closely aligned with major ideologies of domination in the form of industrialization and mechanization of labor; European imperial and colonial hegemony; and, in the present, economic and cultural globalization. Yet, innovation is not a phenomenon of the industrialized West. Archaeologists and historians have innumerable examples of the development, spread, and adoption or rejection of new ideas in myriad past societies. Humans are endlessly creative, and we have manipulated our social structures and technological systems to compete with each other; distinguish ourselves; and explore, understand, and survive our environments, perhaps since even before we were fully human. However, most of our intellectual tools for understanding this process rely on capitalist ways of thinking about invention, technological change, and the adoption of new things (for extensive discussion from the perspective of philosophy of technology, see Feenberg 2010).

As is probably clear by now, I would argue that this intellectual framework is inherently flawed, as it is biased towards understandings of innovation centered on powerful cis-male actors, and their creation and manipulation of consumers. The regular complaints that incremental and process innovations, not to mention innovations in female- and minority-dominated sectors, go unrecognized and uncelebrated reflect the inherent weakness of the capitalist conception of innovation. By contrast, archaeologists regularly work to reconstruct worlds, social structures, and technological systems not controlled by capitalist economic pressures and social dynamics. We are uniquely placed to explore the complexities of innovation, both social and technological, and their shifting form over time.

So, to answer the questions with which I started this chapter: I find innovation to be a particularly important field of research as it is both a driving force in our present society and a conceptual tool used to justify and increase inequality through the imposition of an economic and social value system disguised as a natural process. Archaeology, as both the study of the world before the modern era and a series of methods designed to extrapolate social structures and values from the material residue of daily life, gives us the perspective and the tools to study cultural and technological change in myriad contexts and to propose ways of understanding it that avoid privileging the implicit economic and technological assumptions of the present day.

Notes

1 See C. J. Frieman, "Innovation, continuity and the punctuated temporality of archaeological narratives," in A. McGrath, L. Rademaker, and J. Troy (eds.), *Everywhen: Knowing the Past through Language and Culture* (2023).

2 See, for example, http://ssir.org/topics/category/global_issues (accessed September 12, 2020).

2

Messy narratives/flexible methodologies

How can we understand innovation, given the limits of the archaeological record?

In 1954, Christopher Hawkes, then Professor of European Archaeology at Oxford, published a meditation on archaeological methods based on a presentation he had given a year earlier whilst in the United States as a visiting scholar at Harvard (Hawkes 1954). Although his main thrust was to push back against the then emerging Americanist anthropological archaeology by presenting an alternative view of the discipline as most closely aligned to history, in fact the paper's greatest legacy lay elsewhere. In the course of his discussion, Hawkes articulated the ways the fragmentation and incompleteness of the archaeological record hampered archaeological investigation in a short section usually termed "Hawkes' ladder of inference."

Following Hawkes, the material culture of the past is most accessible to archaeologists and, thus, the techniques by which this material culture is produced are easiest to investigate (this is the lowest rung on the ladder). Next most accessible are the subsistence practices or economic basis of past societies, visible through studies of faunal, floral, and other environmental data. At this stage, Hawkes suggests there is an obvious increase in difficulty as one attempts to identify and explain political and social institutions, such as social hierarchies or gendered power dynamics, from the archaeological record. At the top of the ladder – and well out of reach of most archaeologists – is the question of past peoples' beliefs and spirituality, since these are expressed in ephemeral ways and are often difficult to abstract from the objects used in religious practices. Hawkes was ultimately pessimistic about the ability of archaeologists to use any tools beyond ancient texts to delve into these more social arenas, but archaeologists (chief among them the anthropological archaeologists with whom he was in dialogue) have resisted this pessimism and developed a wide range of analytical and interpretative tools to explore ancient society and past people's beliefs and identities in addition to their technological systems and subsistence practices. It is these tools that we also bring to bear in understanding innovation.

As one might expect, studying innovation as an archaeologist presents a number of challenges. As Hawkes warned, we are forced to work with a limited and fragmented dataset. Moreover, our chronological precision is typically measured in centuries or – at its most accurate – decades, rather than the days, weeks, months, and years over which innovations are developed, tested, and disseminated; and, perhaps most crucially, in almost all cases we lack human informants to describe their own perspective and participation in the innovative processes we are attempting to observe and analyze. Nevertheless, innovation – or, at the very least, technological change – continues to fascinate archaeologists and serve as a primary focus in our attempts to reconstruct past worlds and social structures. At its most basic level, this is nothing more than exploring the lowest rung on Hawkes' ladder: a chronological description of the techniques used to shape material culture in the past.

Consequently, most archaeological research into so-called innovation actually boils down to excited publications (and widely disseminated press releases) about a new earliest date or first-ever artefact. Yet, these earliests and firsts do not in fact tell us anything meaningful about innovation, itself a complex social and technological process that sits rather higher on Hawkes' ladder. Instead, they represent the horizon of archaeological visibility, which, the farther back in time we go, becomes harder and harder to link to the specific moment of past innovation adoption or diffusion. Between natural (and human) taphonomic processes, differential deposition histories, and the mix of randomness and luck in finding an important artefact in any given excavation (which itself represents a minuscule window into an incredibly geographically and temporally vast past world), by the time a technology or phenomenon becomes archaeologically visible it is probably already ubiquitous in the region and period in question. Understandably, this limits our ability to talk about the phenomenon of innovation on the scale of a human life.

Nevertheless, over the last few decades an archaeology of innovation has begun to emerge. Drawing on lines of thought from biology to sociology to linguistics, archaeologists have been re-evaluating technological changes and trying to grapple with the underlying processes and social structures that made them possible. The application of these intellectual tools has, however, been far from uniform. Much of this work is rather ad hoc, stemming from research carried out by single archaeologists (or research teams) investigating a specific case study or phenomenon, but a number of archaeologists working within the wider field of evolutionary archaeology have been making a concerted effort to develop a more generalizable approach to innovation that relies on Darwinian models and metaphors (O'Brien and Shennan 2010; Shennan 2002a).

As a social archaeologist, I am unconvinced by these approaches, so, in this chapter, I explore the power of evolutionary metaphors to explain or model innovation, but I also critique their inbuilt assumptions about the nature of change, as well as their inability to grapple with the complexities of socially constructed technologies and practices. Consequently, I advocate a more human (and hopefully also *humane*) approach to invention, innovation, and social and technological change, relying instead on models developed in the human sciences that make clear that technology is neither separate not separable from society. Thinking critically about how we understand innovation is crucial because changing conceptualizations of innovation, creativity, and agency do not just reflect different approaches to the past, but actually create meaningful differences in how we interpret the archaeological record and evaluate its significance in the present. Nowhere is this clearer than in the rapidly changing picture of the beginnings of agriculture. New data, and a new openness to understanding its implications, are shifting our models of plant domestication away from a simple pattern of expanding diffusion to an almost bewildering complexity of phases of experimentation, adoption, and rejection that lasted for hundreds or even thousands of years.

The growing complexity of early agriculture studies

The history

The story of agriculture is one of the core narratives archaeologists have been refining over the last 150 years. Unsurprisingly, as our suite of methods has expanded, the questions we ask about the origins of domestic agriculture have increased in complexity, and our answers have concomitantly become more detailed and nuanced. Setting aside the hyper-diffusionism of the early twentieth century – models that envisioned, among other things, that all major technologies, including domestication, were invented in Egypt and diffused to all the other continents from there – it is generally accepted that agriculture was independently invented in numerous different places and times. Both the mechanisms of agricultural domestication – how people managed to domesticate wild plants – and the rationale for this behavior have been debated at considerable length. There are a number of reasons archaeologists have focused on this particular area to such an extensive degree.

First, the move to a primarily agricultural subsistence base represents a significant shift away from the reliance on wild resources that characterizes hunter-gatherer ways of life, implying that it would also have had profound effects on how people related to their local landscape, developed their economy, and organized social structures and practices (see, for example, discussions in Diamond 1997a; Price and Bar-Yosef 2011, S168; and Rindos *et al.* 1980). Vere Gordon Childe, early-twentieth-century archaeologist and public intellectual, hypothesized that the invention and adoption of agriculture represented a "revolution" in the Marxian sense: a core shift in both the mode and the relations of production with attendant effects to the cultural and economic superstructure (Childe 1936).

Second, compared to ancient religious beliefs or household power dynamics, it is relatively straightforward to investigate the beginnings and spread of agriculture. The tools and techniques for studying subsistence practices, ancient plant remains, and other environmental data are well established (e.g., papers in Zeder *et al.* 2006), and archaeological methodologies have long been adapted to collect and appropriately process soil samples to preserve this data. Whilst new methods are constantly in development and new means of interpreting environmental data can shift our understanding of the import of these data, there is a deep and rich pool of comparative material, climate models, and methodological expertise upon which geo- and environmental archaeologists can draw in examining the domestication process.

Finally, the beginning of agriculture (and its presumed social, economic, and political impacts) has broad public appeal and is one of the few prehistoric topics regularly taught in secondary schools,[1] suggesting that its continuing tantalization of the archaeological imagination reflects wider intellectual trends regarding what research topics are both compelling and urgent. Certainly, Childe's suggestion that this process was revolutionary and fundamentally affected human social and political formations continues to resonate in the present in both scholarly and public circles (Frieman 2020).

Until very recently, most archaeologists agreed that agriculture was independently invented in only a few locations on earth (Fig. 2.1). In addition to archaeological data, this model was based in the pioneering work of agronomists and biologists, especially the Russian botanist Nikolai Vavilov (1992). Vavilov examined the genetic and phenotypical variation of common domesticates and identified eight centers of plant domestication in the world: Central America, South America, the Mediterranean, southwest Asia, North Africa, Central Asia, South Asia, and East Asia. He later revised this model, recognizing three distinct regions within South America (in the Northern Andes, in the Amazon, and in coastal Chile) and adding North American and southeast Asian/ Indonesian regions.

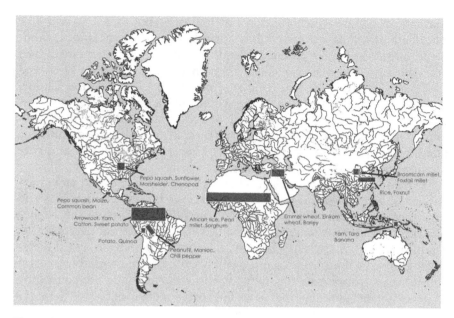

Figure 2.1 A parsimonious model of the locations of specific hearths of plant domestication.

Archaeological research within these regions has sought to identify the chronological phase in which agriculture commenced, and the reason or reasons for its invention and adoption. In line with much of the mid-twentieth-century writing on creativity and innovation in small-scale societies, archaeologists engaged in this research assumed that the act of invention was rare. The underlying principle was the belief that small-scale societies, such as the hunter-gatherer groups who first began experimenting with cultivation over 10,000 years ago, were deeply and innately conservative, and that only major pressure – typically external – could instigate the adoption of new technologies or ways of life (the wider topic of "conservatism" is addressed in Chapter 6) (Tilley 1989, 107–8). Agriculture, apparently a major rupture with past practices, would then seem to be too significant a change to have emerged with any frequency.

From Childe's work onwards, agriculture was assumed to lead to considerable social changes, including increased population sizes, perhaps because of a more reliable or more calorie-dense food base;[2] sedentary lifestyles linked to the eventual development of proto-urban and urban settlements; increasing social complexity leading to the emergence of chiefdom and state societies; and the emergence of specialist and specialized technologies, in part because of the capacity to store agricultural surpluses and in part based in social strategies enacted by newly aggrandizing social elites (Diamond 2002). The underlying assumption here derives from social-evolutionary models of human society, which posit that agriculture as a technology and a new economic system was causal to, or at least correlated in obvious and meaningful ways with, increasing social and political complexity.

It is heavily influenced by the well-known archaeological sequence from the so-called "Fertile Crescent," the birthplace of southwest Asian agriculture and the place where some of the world's early state societies emerged (see, e.g., Bar-Yosef 2011, S175). Here, a period of sedentism and intensive exploitation of select wild resources by hunter-gathers preceded the transition to agriculture. Early agricultural products included pulses and staple grains (wheat and barley), supplemented by some wild resources. The domestication of herd animals followed several generations later, by which time most people were living in permanent settlements, some monumentalized into massive *tells* as new buildings were constructed on top of previous structures. This region has long loomed large in the archaeological imagination, both for its long history of widespread, intensive excavation and for having been the geographic origin of the three major Abrahamic religions. Southwest Asian data on early agriculture is often both the most abundant and the best known; so, consequently, it tends to be generalized more globally, even if this is unwarranted or inappropriate.

Changing methods, emerging interpretations

Identifying and interpreting domesticated plant remains constitute both an art and a science; and extrapolating their significance relies on a much broader and increasingly nuanced understanding of the local environment, climate, and culture in which they were grown. Archaeobotanists studying seeds and other plant remains from archaeological sites have identified a number of changes that happen as a plant is domesticated: the seed, root, or some other organ utilized by humans increases in size; it loses natural dispersal and defense mechanisms, such as long awns or spines; and there are distinct changes in how the plant grows and propagates, favoring the reduction of germination inhibition, and synchronous flowering and ripening (Fuller 2012).

In recent years, plant microfossils and other microscopic remains, including pollen, phytoliths, and starch, have also been used to great effect to identify the presence of domesticated plant species. Starches in particular show great potential, as they vary between domestic and wild species of the same plant and are preserved on lithic implements and human teeth, even in environments where plant macro-remains do not preserve (Piperno 2012). The emerging field of ancient genetics provides further means of identifying and describing domesticated plant species, those traits that indicate deliberate selection and their original source regions (Jones and Brown 2000; Murphy 2007). Alongside these developments, improvements to field methodologies – especially the widespread adoption of soil flotation – and archaeological dating techniques have exponentially increased the available archaeobotanical and environmental data for plant domestication while also establishing a much more precise chronological framework for the beginnings of agriculture around the world.

The results of this burgeoning research point to a much more complex and diffuse pattern of domestication than previously assumed. Individual crop species appear to have been domesticated multiple times, and there are many more points of localized invention or experimentation with cultivation than is imagined in more parsimonious models. Research on rice (genus: *Oryza*), for example, indicates that it was domesticated on three continents: *O. sativa* appears to have been domesticated independently in both the Middle Yangtze and Lower Yangtze valleys from between about 8,000 and 9,000 years ago (Choi *et al.* 2017; Silva *et al.* 2016), and possibly also in northern South Asia from around 4,000 years ago (Bates *et al.* 2017; Fuller 2011); *O. glaberrima* was independently domesticated in one or more regions in West Africa about 2,000–3,000 years ago (Linares 2002); and recent research in South America suggests that *O. Glumaepatula* (and possibly other wild rices of the genus *Oryza*) were being cultivated in Amazonian wetlands from about 4,000 years ago (Hilbert *et al.* 2017).

Even within the well-known and heavily researched southwest Asian "Fertile Crescent," new scientific methods for analyzing, identifying, and dating domestic plant remains suggest that cultivation was widespread, and that domestication events were geographically diffuse, numerous, and protracted (Civáň *et al.* 2013; Fuller *et al.* 2011). This pattern can be seen elsewhere in the world, where new data from genetics, starch analysis, and archaeobotanical research suggest that cultivation – sometimes leading towards domestication, sometimes abandoned prior to the development of a fully agricultural economy – was a common practice; that even the best known cultivars were probably domesticated multiple times and hybridized with each other and with wild progenitors (Fuller *et al.* 2014); and, consequently, that new models have become necessary to understand why people engaged in landscape management to the point of domestication and how this process developed in different areas.

In particular, the idea that there were many communities experimenting with cultivation, rather than the scant eight to twelve advocated by more parsimonious models, suggests we need to alter our image of past innovation in plant management. Indeed, these models reproduce (and perhaps also validate) Eurocentric conceptions of innovation and technology, contrasting the well-known, discrete Eurasian centres of agricultural domestication – in which agriculture is part of a complex Neolithic package that diffuses outwards into marginal areas – with African, South and southeast Asian, and North and South American regions – where plants were domesticated, but in a complex and, at times, diffuse mosaic and did not apparently form part of a larger technological package or social formation (Vrydaghs and Denham 2007, 7–8). In Papua New Guinea, for example, although complex agricultural systems were developed millennia ago, society remained largely egalitarian, with little formal political hierarchization (Denham *et al.* 2003). Bruce D. Smith (2001) highlights this ambiguity through an examination of various societies practicing what he terms "low-level food production": that is, intensive hunting and gathering with some custodianship of wild resources and/or marginal use of cultivated foods. For example, he points to the Paiute tribe, who settled in the wet, low-lying meadows by the Owens River Valley (Great Basin, eastern California), arguing that they occupied a middle ground between classical hunter-gatherers and agriculturalists. They exploited only wild resources, but lived in permanent villages and shaped their marshy environment through irrigation and water management systems. In neither Papua New Guinea nor the Owens River Valley did complex environmental management necessarily lead to specific social or political formations, nor did it necessarily lead to cultivation or intensification.

As discussed in Chapter 1, conflating technological complexity (or even the invention of specific technologies) with social complexity (and, often, superiority) is a hallmark of the European colonial mindset. Removing these assumptions from our models of agricultural innovation means re-evaluating the very idea of domestication, and how we identify it, accord it significance, or position it within narratives of social and technological development. Tim Denham (2009, 2011) has argued persuasively that, in order to understand early agriculture, we must shift our perspective away from teleological economic models. Instead, we should view agriculture as a complex technology made up of plants, people, tools, and culturally meaningful ways of making, shaping, engaging with, and perceiving the local environment. Eurocentric ontologies constructed around a meaningful separation among people, plants, and other living and non-living parts of the environment are not – and, certainly, were not – universal (Descola 2014). Agriculture then becomes an embodied practice and one that is highly culturally contingent, rather than a linear process divorced from other aspects of daily life.

Moreover, despite years of analytical research that demonstrates that plant domestication is more a spectrum than a binary opposition of wild and domestic (e.g., Smith 2001), early-agriculture researchers still insist on hunting for centers in which the earliest examples of "fully domestic" plant species can be found (see, for example, the hand-wringing over Chinese rice domestication in Choi *et al.* 2017). This process itself invalidates and devalues the complex environmental manipulations engaged in by people whose plant cultivation was decoupled from social or political change or failed to diffuse widely, not to mention those whose environmental management does not look like full-fledged agriculture in a European framework.

Over thirty years ago, William Cronon's (2003) germinal environmental history of colonial New England made clear that the Eurocentric conflation of agriculture, sedentism, and hierarchical social systems blinded the early American colonists to the complex landscape management and agricultural technologies used by Native North Americans. Bill Gammage (2013) has made a similar argument for Australia, demonstrating the sophistication of traditional Aboriginal land management prior to colonization and how it shaped the landscape encountered by Europeans. Indeed, Bruce Pascoe (2016), in an award-winning piece of complex ethnohistory, has argued that Aboriginal Australians did indeed cultivate a wide variety of plant species, but that knowledge of the technologies they used were largely lost to history through a mix of European ignorance and intentional erasure of Aboriginal accomplishments.[3] In other words, while early-agriculture studies are still overwhelmingly concerned with dating the earliest centers of biologically full domestication, the emerging body of data – scientific, archaeological, and ethnographic – militates for a view of agricultural innovation that allows for numerous parallel and mutually influencing processes of invention and innovation diffusion with no set trajectory and little fundamental relation to social structure, except in its embodied practice.

Innovation as an evolutionary process

By far the most widely adopted archaeological approach to innovation frames it as a Darwinian process, able to be understood and explicated with tools drawn from the wider field of evolutionary science. The idea that technologies develop in ways that mimic biological evolution is not limited to archaeology (e.g., Basalla 1988; Ziman 2000), but its centrality to archaeological reasoning is, perhaps, unique in an interdisciplinary context. Nineteenth-century archaeology and anthropology both drew inspiration from and developed major theories and methods out of the social evolutionary models prevalent at the time (Díaz-Andreu

García 2007, 369; Riede 2006). European researchers, deeply influenced by evolutionist ideas of progress and teleological narratives of change over time, studied the organization, economies, and technologies of the people they conquered and used them to develop unilinear models of social development in which societies could be ranked on a stratified and hierarchical scale from Savagery to Civilization (the latter of which often looked suspiciously like Victorian London). In some of these rankings, specific technologies were correlated with levels of social evolution, so for example Savages were hunter-gatherers and Barbarians agricultural-ists (as in Table 1.1).

The positivist underpinnings of this form of evolutionism mandated that social and technological development necessarily progressed in a unilinear fashion from simple to complex, with the implication being that increasing technological complexity correlated with increasing social complexity (and, because of the colonialist mindset out of which these ideas grew, value). Clearly, these models were inherently racist and Eurocentric; but, even while the trappings of imperialism have long been shed, evolutionary thinking – now frequently framed as evolution cul-ture theory (ECT) – remains a core part of many archaeological models of change over time.

Robert Dunnell (1980) draws a distinction between the social evolution underpinning a considerable amount of twentieth-century, anglophone anthropological archaeology and the emergence of lines of thought based in scientific or Darwinian evolution. These modern-day evolutionary approaches engage with evolutionary science rather than nineteenth-century social evolution, and are designed to draw on rich datasets and interpretative frameworks from the biological and environ-mental sciences, as well as psychology and cognitive science (for a com-prehensive and current overview of the breadth of the field, see papers in Prentiss 2019). More specifically, they shift archaeological attention away from identifying stages of social development and towards the role that evolutionary principles – such as selection, fitness, and transmis-sion – play in the development and dissemination of human behaviors and cultural practices.

Proponents of the application of evolutionary models to archaeo-logical phenomena particularly note their ability to encompass complex processes and help develop nuanced narratives of human action that take account of both chronological and environmental factors (Walsh *et al.* 2019b). For example, most of the answers to the question of *why* people started to domesticate plants boil down to either "push" or "pull" narratives (Price and Bar-Yosef 2011, S168) – hunter-gatherers are either pushed into beginning to cultivate plants by external pressures or they are pulled to it by its perceived benefits. So, for example, Childe

(1936) and many others believed that climatic change played a large role in pushing hunter-gatherers to begin domesticating. A deteriorating climate in the Younger Dryas (about 12,000 years ago) would have led to increasing aridity and lower temperatures in southwest Asia (and perhaps also East Asia), reducing the abundance of wild food resources and putting pressure on the various groups of people who depended on them, leading to experimentations with cultivation (Bar-Yosef 2011). By contrast, Brian Hayden (1995a) suggests that hunter-gatherers attempting to exert power over kin and neighbors were pulled to accumulate food surpluses for use in social displays of wealth, such as feasting, which would have led naturally to experimentation with agriculture in order to have access to larger food resources. In both models, agriculture is framed as the solution to a problem facing hunter-gatherers.

Botanists and ecologists, by contrast, have developed models that are more neutral as regards human intentionality, such as Rindos *et al.*'s (1980) important suggestion that agriculture emerged as a beneficial symbiosis of humans and the flora in their local ecosystems. This approach resonated with models of adoption promoted by behavioral ecology, a theoretical framework for human behavior that blends evolutionary theory and economic concepts. So, for example, following behavioral ecological models, agriculture emerged because it was an optimal – more efficient and effective – form of food acquisition that would have led agriculturalists to be more fit (in this case meaning reproductively successful rather than individually in better health), and their practices to be transmitted to more offspring (e.g., Boone and Smith 1998).

Within the broad field of archaeological ECT, innovation has been a consistent topic of research (Lyman and O'Brien 1998, 616–17), with attention being focused particularly on two general questions: Why engage in novel behaviors and how do cultural transmission mechanisms affect or disseminate innovations? Answers to the first question are largely the remit of evolutionary ecologists and rely on economic proxies for fitness (Kuhn 2004, 562). That is, they assume that increased access to resources (and especially to a more stable diet) is a valid, measurable stand-in for reproductive success; and, moreover, they explicitly propose that individual humans preferentially engage in activities that will lead to their increased success in resource acquisition. So, following this approach, people engage in novel behaviors when those behaviors will lower the cost of access to resources or otherwise increase their ability to obtain them more efficiently and effectively, especially at a subsistence level. The second question has been taken up by so-called Darwinian, Evolutionary, or "selectionist" theorists, who draw on the principles of biological evolution to explain both technological and cultural change in past populations (at both micro and macro scale: for the latter, see Prentiss and Laue 2019; for the former, see Walsh *et al.* 2019a). The principle of dual inheritance that underlies much of the selectionist intellectual framework posits that culture and genes influence each other and, moreover, that both can be understood as developing and changing in line with similar evolutionary principles (Boyd and Richerson 1985). In other words, genetic models (both mathematical and intellectual) of transmission can be applied to cultural practices or systems; and changing cultural practices can have an impact on the genetic profile of a population. Indeed the recent attention paid to epigenetics – that is, the impact of environmental factors on the expression of genes – supports the idea that culture and biology are linked (Ginsburg *et al.* 2019; Jablonka and Lamb 1995; Jablonka *et al.* 2005).

Stephen Shennan (2000), for example, argues that archaeologists can fruitfully apply the evolutionary principle of *descent with modification* – an inheritance system that recognizes the possibility of imperfect transmission of genetic data because of mutations – to the transmission of behaviors and cultural practices (specifically craft-making processes) over time within a given culture. Based on a brief survey of ethnographic comparisons, he proposes that, like genetic inheritance, cultural transmission is largely vertical (from parent or older generation to child or younger generation) and is a primarily conservative process with few changes emerging from within the culture. Change is largely provoked, he argues, not purely from environmental stimuli but most significantly from demographic stimuli, including large-scale migration. He uses this approach to examine the late fourth millennium BCE transition from the archaeologically identified Pfyn to Horgen Cultures at several well-preserved and finely dated waterlogged settlements in the Alpine region (broadly, modern-day Switzerland, France, and southwest Germany), and argues that changes in way of life and material culture at these sites might be best understood as reflecting periods of abandonment and migration leading to re-occupation by people of varying "lineages of different cultural practices" (Shennan 2000, 833).

Both ecological and selectionist approaches have been widely critiqued – from within archaeology and in cognate disciplines – for their foundational assumptions. Behaviorist or ecological models, for example, by nature of their particular fusion of evolutionary principles and twentieth-century economic concepts, assume not only that individualized economic rationality is an innate feature of human cognition, but also that the efficiency of resource acquisition somehow correlates directly with both social and biological fitness (Bamforth 2002, 438). These are troubling assumptions, as they both universalize and naturalize *Homo economicus*, the mythical, purely rational and self-interested individual whom Sahlins (2013, 161) refers to as a "zombie economic idea that refuses to die," rooting his reality in the biology of our species with little discussion. Indeed, decades of research in anthropology and sociology makes clear not only that economic rationality is a myth in our own culture, but also that it categorically does not exist in small-scale societies. Bettinger and Eerkens (1999) address this, to some degree, within a selectionist context in their study of bow-and-arrow technology in the Great Basin of California and Nevada. They demonstrate that some forms of cultural transmission operating within this region pre-empted or forbade individual experimentation, resulting in less optimal and efficient arrowheads, but that these very restrictions may have developed out of "group-beneficial cooperative behaviours" that would reflect the effect of selective forces operating at the group rather than the individual level. In other words, individually irrational behaviors can be seen to result from cultural practices that benefit group cohesion and contribute to the ability of the group to thrive, even if some group members are less efficient hunters.

This tension between individual and group fitness can be resolved only when we accept the anthropological supposition that the individual self is a culturally and temporally mediated invention rather than a universal form of self-identification (Fowler 2004; Strathern 1988). This would seem to contradict the frequently articulated concern that evolutionary models disregard or diminish the role of human agency, but in fact what is argued is not that ECT fails to recognize individual choice (it manifestly does not), but that it typically recasts creativity – and particularly the creative manipulation of social and environmental structures – as a response to externally provoked necessities (be they environmental or demographic in origin, depending on whether one's approach is ecological or selectionist).

We might look again to agriculture for an example of how this plays out in archaeological explanation. The wealth of research on the spread of agriculture beyond its (proliferating) regions of invention makes clear that cultivation was not adopted in isolation, but often formed a complex "package" of social, political, and economic practices. It was not a purely environmental phenomenon – nor a purely subsistence-based one – but was also intimately linked to whole ways of life, from diet and clothing to social relations, that, depending on region, included shifting gestures and practices, the exploitation of new raw materials and technologies, changing social identities, novel rituals and religious rites, specific language groups, architectural styles, and patterns of mobility or sedentism.

In some parts of the world, population movement (specifically in the form of the slower *demic diffusion* rather than the invasions imagined in an earlier era – see Chapter 5 for a fuller discussion) seems to have been responsible for the spread of agriculture – that is, people who came from agricultural societies brought it with them as they moved into new areas (Ammerman and Cavalli-Sforza 1984, 1973). Recent genetic studies of ancient human remains seem to support this hypothesis for the first agriculturalists in the central and western Mediterranean (e.g., Valdiosera *et al.* 2018), whilst, in others, hunter-gatherer groups living adjacent to agriculturalists eventually adopted agriculture, often after a long period of co-existence.

In southern Scandinavia in the fifth millennium BCE, a combination of these processes seems to have occurred. People known archaeologically as the Ertebølle Culture developed a unique social structure and settlement pattern in Denmark and southern Sweden – socially at the interface of Baltic/eastern European hunter-gatherer groups and the agricultural Linearbandkeramik groups to the south. Their economy and ceramic tradition clearly link them to their eastern Baltic neighbors (Piezonka 2012), but they were also obviously in regular contact with the agricultural communities to their south, as most famously evidenced by their use of stone axes known as shoe-last adzes, which originated in central Europe (Fischer 1982). Ertebølle sites have also occasionally yielded sparse evidence for small numbers of domesticated plants and animals, including dairy products (Courel *et al.* 2020), suggesting not only that these too were traded between agriculturalists and hunter-gatherers, but that the hunter-gatherers themselves were well aware of Neolithic ways of life and subsistence practices, and actively chose not to adopt them (Klassen 2002).

In fact, when agriculture came to southern Scandinavia around 4,000 BCE, it came rapidly and in a complex package, along with long houses, megalithic architecture, and the material culture and practices we term the Funnel Beaker Culture, including the use of the earliest copper in northern and western Europe (Roberts and Frieman 2015). While it has long been argued whether this represented population movement or local adoption (e.g., Johansen 2006), recent aDNA research suggests that genetic turnover seems to have occurred by the middle of the fourth millennium BCE (Sánchez-Quinto *et al.* 2019). However, this does not mean that Ertebølle populations were killed or evicted by incoming agriculturalists, but that more and more intimate engagement with southern agriculturalist populations was a part of the shift to agriculture in this area. Indeed, settlement pattern, lithic technology, and diet all show significant signs of continuity from the 'Mesolithic' fifth millennium BCE to the 'Neolithic' fourth millennium BCE, suggesting a process of negotiation through which technological and social innovations were selectively integrated into the local ways of life alongside new people (Craig *et al.* 2011; Gron and Sørensen 2018). In this case, neither selectionist nor ecological models can adequately explain this constellation of changing practices without consideration of the complicated and historically rooted social practices and relationships of the people who engaged in them.

Approaches to technological change rooted in behavioral archaeology seem to offer a compromise between the naturalistic innovation models promulgated by evolutionary archaeologists and more socially oriented ideas (Schiffer 2011; Schiffer and Skibo 1987). Behavioral archaeologists center human action and develop a model of technological change in which technology is itself acknowledged to be a complex bundle of processes, practices, and bodies of knowledge; however, technological function (whether physical or social) is the core area where innovation is deemed to occur. As external necessities (stemming from lifeway or social organization) require changes in behavior, the functions of tools and other artefacts are altered, and innovation occurs. So, for example, as a new crop becomes a staple, different tools are developed to cultivate, harvest, and process it, and these may take on ritual or social functions as well. Thus, innovation is still conceptualized as responding to needs – often subsistence-based ones – but the complex formulation of technology allows for human creativity, social practice, and relationships to influence the form and direction these innovations take. Michael B. Schiffer (1996) identifies a number of commonalities between ECT and behavioral archaeology, including their shared central focus on variation over time and their attempts to understand and explain knowledge transmission. Nevertheless, he strongly questions the ability of broad evolutionary models to cope with the sheer variability of the human-scale, fragmented archaeological record.

Beyond narrow disciplinary critiques, the wider implications of ECT-based models also deserve scrutiny. To draw on feminist anthropologist Susan McKinnon's (2005a, b) critique of evolutionary psychology, this approach enacts bad science to model the evolution of behavior by studying a narrow section of contemporary society – most evolutionary psychology studies utilize upper-middle-class, anglophone university students as subjects – and then projects culturally specific values, beliefs, and practices back onto the past, normalizing and universalizing them in the process. McKinnon describes this research as ideologically neo-liberal and largely teleological: the contemporary attitudes that are identified in evolutionary psychology studies are considered adaptive because they exist in the present, and a rationale for their benefit to evolutionary fitness (generally equated with radical self-interest) is detailed as proof. Often, this rationale includes reference to one or more case studies of cherry-picked ethnographic or cross-species analogies. These "just-so stories" have been widely employed by evolutionary archaeologists as well, as support for their interpretations of technological change and cultural transmission. For example, Laland and Reader (2010) offer four separate case studies of innovation in non-human animals (monkeys, starlings, guppies, and rats, respectively) and use the observed behaviors to generalize that innovators are likely to be low-status, hungry, and small (since animal innovation is largely about struggling for food sources) and that sex differences in innovative behavior are best understood as part of wider mate choice or parental investment strategies.

On the surface these are rather mundane observations, but, following McKinnon (2005a, 125–7), we should be wary of applying them to human populations. She argues that this kind of narrative implies the existence of a sort of psychic unity of all species: that is, animal behavior is anthropomorphized, being given human-like motivations and relationships, while humans are made animalistic, their interactions, practices, and engagements shorn of culture and attributed only to biological imperatives.[4] A stronger case might be made for comparing innovative activity between humans and non-human primates. Indeed, Bandini and Harrison's (2020) review of the literature for chimpanzee innovation highlights a lack of consistency or predictability regarding those innovations that "catch on" and those that are not adopted even when they might be beneficial to survival. They suggest that, as is the case among humans, "individual characteristics and personality" (28) play key roles in determining whether a chimpanzee becomes an innovator or adopts an innovation. Having based their model of innovation only within the animal-behavior and evolutionary literature, they are unable to account for social aspects of innovative practices among primates.

A more fundamental flaw, however, is that evolutionary models do not and fundamentally cannot account for non-reproductive individuals and non-heterosexual behavior. As Bamforth (2002, 438) notes, models drawn from ECT – for example, those concerning foraging efficiency – assume that a stable diet is equivalent to reproductive fitness, but do not address those whose diet is sound but who do not reproduce – a population that might include almost anyone, but certainly does include those whose sexual activities involve only same-sex partners. Indeed, Riede *et al.* (2018) suggest sexual maturity inhibits creative and experimental behavior – most purely expressed in child's play – since it poses a risk to successful reproduction. Similarly, observations about sex differences in innovative behavior account only for binary male and female sexes, fully effacing the existence of intersex individuals, not to mention the complex spectrum of gendered identities and presentations.

In essence, then, explanations of cultural transmission, innovation, and behavior drawn from ECT models are not only innately anachronistic, they (almost always unintentionally) de-humanize and de-culture human behavior in ways that reproduce colonial narratives equating economic activity (e.g., innovation) with cultural sophistication. Further, they embed heteronormative and heterosexist assumptions that flatten the rich and complicated human experience of identity and relationships, and exclude non-binary people and same-sex sexual practices. We cannot begin to understand the human experience (in the past or the present) if we take such a reductive view. In the words of another well-known feminist theorist, Simone de Beauvoir (2009, 62), "Humanity is not an animal species: it is a historical reality. Human society is an anti-physis: it does not passively submit to the presence of nature, but rather appropriates it."

It is for these reasons that I choose largely to sidestep evolutionary principles in my discussion of innovation. I will return to some concepts, such as cultural transmission (Chapter 5) and the evolution of cognition related to creativity (Chapter 7) later, but I distrust the foundational assumptions of the broader ECT agenda. In particular, in exploring how and why people innovated or rejected innovations in the past, I feel strongly that the particular social dynamics from which new ideas emerged and in which they spread is crucially important, and is obfuscated by the generalizing explanations of ECT. As stated in Chapter 1, I do not believe that innovation as a process can be separated from people's relationships with each other, with their wider communities, and with the material world. Thus, I suggest that we look to a more social and less sterilized approach to understanding innovation.

Towards a social archaeology of innovation

I would contend that innovation is not an event, but a multi-stranded and drawn-out process that plays out in the social sphere, as much as the economic or technological ones (if, of course, these can even be disentangled). Where evolutionary perspectives model innovation at a group level, subsuming individual choices, practices, and relationships in the biological, lineage, or community, a social perspective starts at the human scale and centers individual attitudes. Moreover, in contrast to the evolutionary literature, social models (and particularly ethnographic research) indicate that innovation is not only frequent, but also continuous. Homer Barnett (1953) defined innovation as any new thing, tangible or abstract, including both inventions and re-interpretations. This extraordinarily broad definition might, at first glance, seem to imply innovation is simply too ubiquitous to study; but I would suggest it instead allows us to re-frame our questions in ways that not only suit the fragmentation of the archaeological record, but also allow for the many and variable ways people engage (and engaged) with each other, with technology, and with the wider world. In other words, Barnett's catholic approach to innovation gives us a means of stepping around the limitations of parsimonious models and their inbuilt, colonialist conceptions of cultural conservatism. This is a particularly important way to conceive of innovation when we consider the small-scale societies of deep prehistory, which certainly looked immensely different from ours, and yet are often uncritically discussed as simple and unchanging. Moreover, the very ubiquity of innovative behavior reflects the constantly changing social and technological environments in which people are enmeshed (Rabey 1989). Thus, as discussed in Chapter 1, to develop a study of innovation, I would start from the following three premises:

- innovations are embedded in and arise from complex interrelations of individuals (not all of whom are human), corporate groups, technologies, technological systems, and specific geographies;
- innovations are and always have been frequent;
- innovations have histories that affect their development, spread, and perception.

Certainly, many of these assumptions underlie classic narratives of innovation and innovation-diffusion developed in the wider human sciences. It is impossible to discuss innovation without recounting Everett Rogers' considerable impact on the wider field of study – and on the popular imagination (Rogers 1962). Rogers was among the first to highlight the considerable importance of social relationships on the spread (or rejection) of innovations, focusing on the role of influential community members in the spread of innovations within their social networks. He modeled the success of an innovation not by its ubiquity or its economic impact, but by the rate of its uptake by different ideal types of adopters: innovators, early adopters, the majority, and the laggards (Fig. 2.2). Rogers' model has been applied to historic as well as contemporary case studies, and archaeologists and historians have found similar processes of adoption of specific types of architecture or material culture in the past. Certainly, Dethlefson and Deetz's (1966) well-known battleship curves, which illustrate the slow adoption and eventual replacement of motifs on colonial American grave stones, illustrate this clearly.

To Rogers, the determining factor for the successful spread of an innovation was not just its functionality, or how it resolved or filled a perceived need, but whether the wider social group – and particularly influential leaders within it – found it acceptable. Even the idea of necessity is unpacked, as Rogers (1962, 166) points out that the decision that one needs something can be prompted or even created

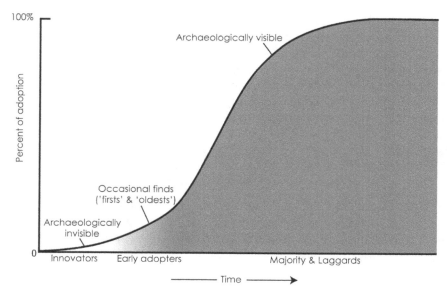

Figure 2.2   The S-curve of innovation adoption over time, with archaeological visibility indicated.

whole-cloth by exposure to an innovation or by an influential "change agent" working to persuade others to adopt one. He broke the process of innovation into five sequential stages, each dependent on communication between adopters and innovators: Knowledge, Persuasion, Decision, Implementation, and Confirmation (Fig. 2.3). Nevertheless, although Rogers acknowledged that innovations are rejected just as often (or more so) than they are adopted, the model he presented is strikingly linear, with little space for parallel or convergent developments, changing practices, small alterations, or partial failures.

Complicating this linearity became a major agenda for a cluster of researchers in the 1980s and 1990s who were intent on developing a model of technology that recognized the significance of the social, not as an external influence, but as a fundamental and intrinsic part (MacKenzie and Wajcman 1985; Pinch and Bijker 1987). Any given technology, they argued, was socially constructed: that is, it emerged from complex interactions among social groups, value systems, other technological systems, perceived needs, perceived risks, etc.

Bijker (1995) famously illustrated this complexity through a discussion of the development of the safety bicycle – the modern bicycle many of us ride today, with two similarly sized wheels encased in rubber tires, and its seat and handlebars set at more or less the same height. He suggested that we have a tendency to describe innovations

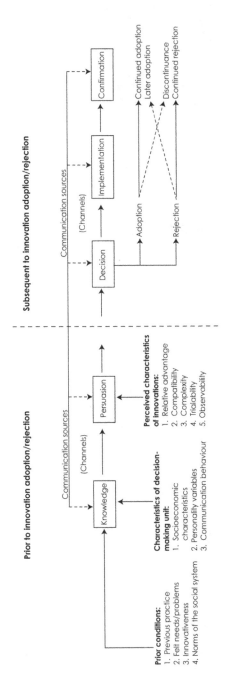

Figure 2.3 The innovation–decision process.

as linear and logical in a post hoc frame, when in actual fact they were messy and emerged from negotiation, experimentation, and successive periods of adoption and failure. So, most histories of the bicycle draw a straight line from the high-wheeled penny-farthing to the lower-wheeled Lawson's Bicyclette to the safety bicycle, emphasising similarity of form, function, etc. (Fig. 2.4a). However, looking at nineteenth-century dialogues about bicycle usage and at the myriad failed bicycle forms – including those that were extremely successful for long phases – makes clear that not only was the invention of the safety bicycle not a logical development based on a slow evolution in bicycle form, neither was it necessarily the only form of bicycle that could meet the needs of cyclists. Instead, he elaborates a model based around different social groups who used bicycles (women, sports cyclists, elderly men, touring cyclists, etc.), problems they perceived with different bicycle models (cycling in a dress, low speed, perceived safety, vibration, etc.), and the solutions various cycle forms offer to these problems (body shape, size of wheels,

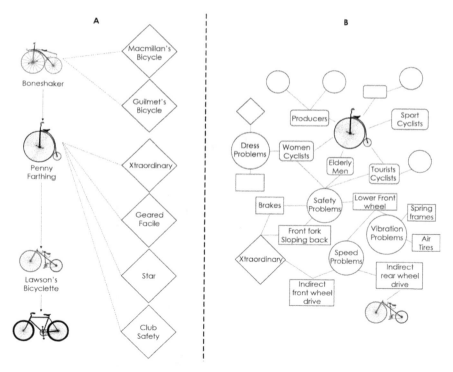

Figure 2.4   (a) A linear, post hoc model of the development of the modern safety bicycle; (b) a constructivist model of the relationship and many competing influences on the penny-farthing and Lawson's Bicyclette.

brakes, air tires, etc.) (Fig. 2.4b). Myriad forms of bicycle, including a number with smaller wheels and more stable bodies (like the modern, or safety, bicycle) were produced contemporaneously, and appealed to certain social groups, but not others, because they offered solutions to specific problems.

Bijker argues that the apparently linear trajectories of innovations traced in other research are, in fact, a post hoc smoothing of this complex mess of social groups, problems, and solutions that materializes from processes he refers to as stabilization and closure – essentially, the emergence and acceptance of social consensus around a new idea or technology. With closure, the flexibility and experimentation that developed out of competing social groups with different problems and conceptions dissipates. So, the modern bicycle not only outlives its older and equally prominent contemporaries, but also comes to look like an obvious solution to all of the problems that shaped the myriad of competing cycle forms in the nineteenth century, even though its own development was haphazard and circuitous.

Innovation, of course, has both temporal and spatial dimensions as well. The complex relationships within which innovations are embedded and out of which they emerge also have histories of their own, which affect the nature, conformation, and success of any innovative practice or technology; and these histories adhere to and are connected deeply with specific geographies. Within the broad field of organizational studies, this temporal quality has been variously examined through the related (but not equivalent) concepts of *path dependence* and *imprinting*. Path-dependence models examine the process by which technological system, organization, or innovation develops through time in line with historically contingent practices, expectations, and patterns of knowledge transfer (Martin and Simmie 2008; Mouritsen and Dechow 2001). As a conceptual framework, it dovetails neatly with Bijker's model of socially constructed technologies: whereas Bijker is interested in the formation of a technology up to the point of stabilization and, ultimately, closure, path-dependence models build beyond this point to examine how the consensus reached during these phases (termed "path creation" and "path lock-in," respectively) carries into the future, affecting both the trajectory and outcomes of subsequent developments. In other words, while Bijker was interested in how the safety bicycle was invented and became the dominant form of bicycle, a path-dependence model would explore how its dominance continues to shape the form of pedal-driven technologies.

By contrast, imprinting is focused more on detailing the social, geographic, technological, and environmental contexts that shaped a given organization at its point of origin, and that are believed to continue to shape it as it develops (Johnson 2007; Marquis and Tilcsik 2013). Similarly, because innovations spread through human networks, their dissemination and patterns of adoption also develop spatial patterns that then impact the spread of future innovations. Torsten Hägerstrand's (1966, 1967) germinal work in this area emphasized that the diffusion of innovations could equally be modeled geographically, with the S-curve here representing the widespread adoption of a novel practice or technology from a small, geographically bounded cluster outwards through adjacent or allied "neighborhoods" and to the wider region. Following his models, routes of innovation transfer persist through time, since they follow established connections between individuals and communities.

Within archaeology, it is well understood that patterns of behavior, technological choice, social organization, and subsistence practice can persist over time, creating meaningful distinctions in the ways different groups of people engaged with each other and inhabited the world. This is, in fact, the basis of the idea of archaeological "cultures" developed in the nineteenth century and still a key part of our explanatory toolkit. Childe (1929, v–vi), who brought the concept into the anglophone literature, described archaeological cultures as "pots, implements, ornaments, burial rites and house forms – constantly recurring together" in a bounded region and remaining largely unchanged over time until altered by an external stimulus, typically a migratory event. This perspective, common throughout the nineteenth and earlier twentieth century, that small-scale and ancient societies were largely conservative, uncreative, and resistant to innovation, emphasized continuity over the long sweep of history, and helped make sense of an archaeological record in which individual objects and materials changed, but patterns of practice within bounded regions and lasting generations could still be discerned.

In the past few decades, archaeological research into change and persistence over time has built on and incorporated the intellectual framework developed by Annaliste historians (Bintliff 1991). This school of history seeks to move beyond the study of specific events and towards a conception of history that situates these particularities within their geographic and historical contexts. So, for example, Ames (1991) takes inspiration from the Annalistes to examine the development of social complexity among First Nations people of North America's Pacific northwest over the past 11,000 years. He tacks back and forth between the ethnographic record – a temporally restricted documentation of social formation – and a variety of archaeological data (e.g., crafts, houses, and food resources) that provide information about medium-term social developments, as well as climatic and demographic data that offer insight into longer-term patterns of landscape occupation. Through this approach, he is able to develop a nuanced discussion of the history of hunter-gatherer people in this region that emphasizes the dynamism of cultural practices and the heterogeneity of ways of life at different points in time. Although he notes that artistic styles and art making seem to be deeply rooted in the Pacific northwest, other elements – from household size to architecture styles – are more temporally and spatially contingent.

Historicizing models such as these hold considerable value for innovation studies by humanizing and contextualizing innovation adoption. For example, archaeologists have spent generations studying the adoption of metal objects and metallurgy, but only in recent years has the slow adoption process been analyzed. Peter Bray (2012) re-analyzed decades of archaeometallurgical analyses of the earliest metal objects in Britain and Ireland and was able to demonstrate that it took generations for some of the most significant physical properties of metal – its plasticity and recyclability – to be fully appreciated and understood. Instead, his research suggests that the earliest iterations of metal tools were treated much like the products of better-understood reductive technologies (i.e., stone-working, woodworking, etc.). Marie Louise Stig Sørensen (1989) identified a similar pattern in the earliest use of iron in northern Europe. She identified a number of very early instances of iron working in which the iron was worked and utilized like bronze (and, in the case of ornamentation, like a less attractive and perhaps less valuable bronze), instead of a raw material with its own distinct properties.

In a recent critical overview of scale in archaeological thought, Robb and Pauketat (2013) argue that understanding change and telling stories of deep history relies on multiscalar approaches that see relation extending from the small-scale lived experience to broader historical and environmental trends. This approach forgoes causation for the construction of thick narrative building that allows for – and indeed centers – complexity. As I argue in Chapter 6, such a perspective has the potential to highlight persistence and resistance, practices that are all too often ignored in our change-oriented approach to past technologies.

Working from fragmented data

As archaeologists, our primary role is to study the material record of past people's activities, to determine what it comprises and to which period it can be dated, as well as to begin to ask why people did those things. Yet we are severely hampered in this interpretative mission by the very archaeological record we set out to study. It is not complete; and, moreover, it is constantly diminishing in completeness because of the activities of archaeologists, developers, industrialists, agriculturalists, and the natural world (Nickens 1991). Despite the increasing intensity of industrialization and population growth around the world over the last century, the destruction of the archaeological record is hardly a recent development. Randsborg (1998) has detailed the evidence for Danish Bronze Age barrows having been tunneled into, coffins opened, and their valuable contents removed during the Bronze Age – probably within living memory of their construction. Similarly, Tim Taylor (1999) estimates that, of all the metal objects deposited during the southeast European Copper Age, only a minuscule proportion has remained untouched and intact to the present day. Even when material survives hundreds or thousands of years in the ground, by the time it is recovered by archaeologists it is rarely perfectly preserved: organic materials decay, and natural, taphonomic processes break up contexts and muddle associations. A whole field of research into the formation processes that go into production of the archaeological record as we find it was developed by Michael Schiffer in the 1970s and 1980s (Schiffer 1983, 1987). He focuses on the intersection of natural processes ("n-transforms") and anthropogenic or cultural processes ("c-transforms") in the variable destruction and preservation of material in the ground that create the deposits and sites we excavate. What has been preserved, moreover, is often hard to assign to time periods more precise than centuries or, if we are truly lucky, generations, rather than the much more human scale of days, weeks, or years (Foxhall 2000).

Understanding innovation in the past requires us to take these limitations into account and build them into our methodologies. Innovation happens at several geographic and spatial scales: that of the individual and her momentary choice to try or reject a novel thing or idea; that of her experiences with it during repeated trials; that of her family or nearby community as they encounter the innovation through her; that of her community as they become habituated to the innovation, or as conflict emerges over whether or not to accept it more widely; that of the wider region as they encounter and grapple with the innovation; that of the generation of people who recognize the innovation as novel, but are far enough removed from its introduction to have already become habituated to it; that of the supra-region where patterns of innovation adoption and rejection can be mapped; that of the multiple successive generations who continue to use or eventually discard the innovation (Fig. 2.5).

All but the last of these are effectively invisible to archaeologists working on regions and periods without written or oral histories – or, to put it in other words, by the time an innovation is archaeologically visible, it is likely to be well on its way to widespread adoption (Fig. 2.2). So, a first step in studying the innovation process is determining the type and quality of data, its variability over time and space, and its

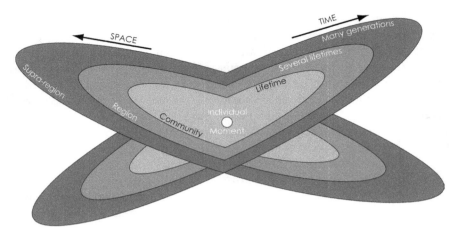

Figure 2.5 The intersecting spatial and temporal scales of innovation.

chronological resolution. Tacking back and forth between different scales – from the moment to the archaeological period, from the site to the supra-region – contextualizes the available data and compensates for the lacunae (Frieman 2012c). This process embeds the spatial and temporal context necessary to understand the process of innovation within our analysis of the data themselves, but also allows for multiple channels of interpretation to co-exist and co-inform each other. A grain of domestic rice, for example, might give us insight into the agricultural practices of a small community at a specific point in time (the stratigraphic layer of the archaeological site in which it was found), but it is also a crucial data point for our ongoing understanding of the longer-term and wider-spread process of domestication in its region (and supra-region) of origin. Neither of these scales tells the complete story of the innovation process that led to rice domestication, but both allow overlapping and complementary narratives of domestication to be drawn out so that the human-scale experience of this process is not lost and the post hoc appearance of linear progress can be minimized.

A key point, and perhaps one that seems obvious on the surface, is that the more data we have, the more individual stories can be woven together into thicker narratives, and the richer our understanding of the innovation process can be. Of particular importance is recognizing that a given innovation does not exist within a single technological or social domain, but inevitably crosses several. Agriculture, for example, is a complex techno-system made up of a variety of tools (each in its own right a nested set of raw materials, bodies of knowledge, experience, and gestures of production and use), practices, ways of managing the environment, specific forms of social organization, ritual practices, cosmological structures, and economic considerations, not to mention the variety of plants under cultivation. Innovations in this techno-system inevitably do not just affect or emerge from those plants. Thus, the innovation itself must be unpacked before it can be studied, and the variety of elements that it enfolds can be targeted for different sorts of investigation. For example, as I mentioned above, Denham (2009) proposed a practice-centered approach to understanding agriculture in the highlands of Papua New Guinea. Instead of focusing on the plants and their biological domestication stages, he unpacks the various types of cultivation, and the tools and gestures that form part of these, in order to draw attention both to the variety of agricultural transformations present within this region and to the continuities in practice that persist through time.

Material culture specialists are already well aware that understanding a given tool or technology requires a much wider appreciation of the technological and material context in which it was produced (McGovern and Notis 1989; Shimada 2007). Metal, for example, cannot be produced without ceramic crucibles and sophisticated pyrotechnology in addition to mining tools and technologies at the very least. A cross-craft perspective treats technologies as interwoven, allowing for a similar sort of tacking back and forth between gestures, raw material studies, and levels of scientific and morphological analysis as we practice at spatial and temporal scales. In small-scale societies, we rarely find crafters who specialize in a single material or practice. Instead, specialization operated at numerous scales – from the amateur to the artisan – and inevitably implicated or involved cross-craft knowledge and experience, and nested specializations. So, for example, excavation of a third-millennium BCE pottery workshop at the Harappan site of Nausharo (province of Baluchistan, Pakistan) yielded a group of flint blades, specially manufactured with copper tools and then used to finish clay vessels (Méry et al. 2006).

Obviously, this approach requires rather more time and a greater number of areas of expertise, but I would argue it also provides a richer, more nuanced, and less fragmented view of past people and their practices. Over thirty years ago, Ian Hodder (1986) was calling for the development of *contextual archaeology*, an approach grounded in the appreciation and examination of the social and technological spheres in which the material we excavate was developed, used, and deposited. Rogers and Shoemaker (1971) outlined a series of social elements that are inherent to the innovation process (Fig. 2.3); whilst these are not all accessible to archaeologists (particularly those working in remote periods), many of them are. We can (and do) study patterns of exchange and value creation. The manufacture and functional capacity of given technologies is a well-established archaeological specialism. Much of our research is dedicated to understanding elements of social organization, including norms and networks of contact and communication.

Ultimately, a social archaeology of innovation is not so different from good archaeology. We recognize the limits in our data, we use the heuristic of the innovation process and the understanding that technologies are complex and complexly social to identify complementary data sources, and we juxtapose narratives of innovation at different scales and emerging from different archaeological data to build thick stories about past practice. Not all these narratives will align – some will dovetail perfectly, others will contradict each other – but that only adds to the richness of the models we create.

Notes

1 Certainly, the "Neolithic Revolution" was part of the world history program at my American state high school and represents the first pedagogical contact I had with archaeological narratives beyond the classical world and Egypt.

2 Whilst there remains considerable debate about the impact of early agriculture, this particular concept has been roundly dismissed: agriculturalists (particularly in the Neolithic) are less healthy, have worse teeth, are shorter, and die of more gruesome diseases than their hunter-gatherer neighbors and ancestors (Larsen 1995). Even today, with all of our technological advances and abilities to predict and work around weather events, farming is a hard way of life, and this must have been doubly true for prehistoric subsistence farmers. By contrast, careful studies of diet and time-allocation among contemporary hunter-gatherer societies make clear that they spend less time, energy, and worry on food acquisition than their agriculturalist neighbors, and far more on recreation and leisure (Binford 1968).

3 I want to highlight the cultural importance that Pascoe's book has had in Australia since its publication. Aside from the numerous book awards for which it has been nominated and that it has won (including the 2016 Indigenous Writers' Prize in the NSW Premier's Literary Awards), it is a regular topic of discussion on Indigenous social media, where its thesis is held up as an example of the sophistication and dignity of Aboriginal Australian cultures and peoples, and it has recently inspired an acclaimed adaptation by Aboriginal and Torres Strait Islander contemporary dance troupe Bangarra. Nevertheless, thanks to the persistence of colonialist and Eurocentric conceptions of technology and social value, in the popular imagination agriculture is still perceived to be a major technological accomplishment and one that implies a level of sophistication, knowledge, and respect owed. Emphasis on innovative centers and laggard peripheries in our models reinforces these conceptions and unintentionally confirms the fallacious assumption that non-agricultural people were less sophisticated, intelligent, or creative than their agriculturalist neighbors. A perhaps unfortunate side effect of Pascoe's and other work on Aboriginal and Native American agriculture is that, in order to access the social value accorded to the innovation

discourse around agriculture, this research tends inadvertently to gloss over the spectrum of cultivation behaviors in its need to identify First Nations people as sophisticated agriculturalists rather than non-agricultural hunter-gatherers.

4   It is worth noting that feminists in particular have long struggled with the nature–culture duality. The idea of nature is historically feminized, and female bodies similarly coded as closer to nature, particularly because of the visceral experience of pregnancy and childbirth. This occurs in both feminist and non-feminist writing, but, in each, it relies on particularly abrasive and gender-essentialist conceptions of womanhood being connected to female biology and biological processes. In other words, nature (female) and culture (male) are not just opposed, but are historically unequal. It is for this reason that I find naturalizing metaphors that efface culture deeply suspect, as they seem perfectly designed to enact and replicate contemporary inequalities while maintaining an effect of scientific neutrality.

3

Invention as process

How do innovations happen? How are they invented?

Invention seems as if it should be particularly difficult to access in the archaeological record, since it is believed, almost by definition, to be a singular event. We excel at delineating and disambiguating the histories of patterns of activity that extend through time and across space and that, necessarily, involve multiple if not scores of individual people and activities. Spotting a singular experiment or an initial conjunction of practices is rare, given the limits of the archaeological record discussed above, even more so because those momentous (emphasis here on *moment*) events are largely ephemeral – existing within the head of the inventor or emerging from an already well-established toolkit, material repertoire, or pattern of practices.

Where archaeologists have access to this sort of temporally constrained and transient activity, it is because the sites themselves – and the processes that created them – are extraordinary. Shipwrecks such as the Uluburun, discovered off the Turkish coast in 1982, give us a window into the lives of their final crews and the progress of their last voyages (Pulak 2012). From their preserved timbers and undisturbed cargo we can date them; determine their routes and home ports; and begin to identify the numbers, status, and roles of individual crew members. Time-capsule sites such as Pompeii and the recently excavated British Bronze Age settlement at Must Farm, Cambridgeshire, give an even stronger impression of a moment or series of moments frozen in time (Knight *et al.* 2019). These sites were destroyed by sudden calamities – a volcanic eruption in 79 CE at Pompeii, a fire around the ninth century BCE at Must Farm – but the nature of the destruction events led to exquisite preservation. So, in the archaeological assemblages at each site, we are confronted with the moment of destruction itself. At Pompeii, archaeologists have used plaster to fill voids in the ash, revealing the bodies of individuals who died during the eruption; at Must Farm we see, among other things, a ceramic vessel full of carbonized food with the wooden stirring spoon still stuck within.

However, sites like these are exceptionally rare, so hoping for archaeological evidence of singular acts of invention is likely futile. That said, from an archaeological perspective, it is worth our time to question whether *invention* is simply a momentary conjunction of person, place, thing, and concept, rather than a more extensive and culturally embedded process. Consequently, this chapter will explore the concept of invention – both as a creative act and as part of larger technological systems – to suggest that, like innovation, it emerges from and exists within complex relationships between individuals, technological systems, and wider social networks.

I will discuss how or whether invention can be distinguished within the broader innovation process, and examine the role of replication and mimesis. Then, in light of Killick and Fenn's (2012) concerns that most archaeological narratives of invention and innovation are innately presentist, I will attempt to delineate how invention might occur and be experienced by people in the past. This discussion is situated against the case study of the complex and rich field of early-metal studies, where new and increasingly precise scientific methods are giving us ever more nuanced insight into the technological processes developed and used by prehistoric people.

Inventing metallurgy in Eurasia

A number of innovations are bundled together under the umbrella of "the invention of metallurgy" (Killick and Fenn 2012, 563–4). To make metal from ore, people had first to realize that some stones would, should the right amount of heat be applied, turn to liquid and solidify as metal. While native copper was certainly among the earliest metallic minerals to be collected and worked, a variety of colorful ores were rapidly ground and smelted into copper as well (Roberts *et al.* 2009). Ores had to be recognized, and the appropriate grinding and smelting processes needed to be developed to transform them. Smelting implies both carefully controlled pyrotechnology – an extension of an already widespread technology used to produce ceramics – and new types of ceramic that could withstand the heat of fires needed to turn stone to metal and contain molten copper without fragmenting (Roberts and Radivojević 2015). Molds, first open forms in stone and later closed stone-and-ceramic moulds, as well as one-use molds utilized in lost-wax casting techniques, were invented, and whole new toolkits for cold working and revitalizing finished objects would have been assembled. While the raw materials and technical processes of many of these ancillary technologies would have been familiar, their usage, forms, and juxtaposition were innovations in their own right (Radivojević 2015).

Beyond technological innovations, numerous social, economic, and material changes were also bound up in the invention and spread of metallurgy. The physical properties of metal are unique among prehistoric materials, introducing concepts of plasticity and recyclability to people's understanding of their material world. That is, in contrast to other available and widely used materials such as stone, ceramic, or fibre, copper (and other metals) could be bent, twisted, or warped and, through the application of heat, be returned to their original form (or even to a totally different form) without noticeable loss of mass or the retention of scars from their damage or re-working. Some archaeologists have suggested that metallurgy would have been a uniquely challenging technology, requiring a level of technical know-how (embodied experience) and knowledge (intellectual understanding) that could only be developed by full-time or near-full-time specialists (Radivojević and Rehren 2016).[1] If so, we might imagine that systems of knowledge transmission and new political and economic structures would have had to be invented in order for metalworking to take hold and become viable. Moreover, after the invention of metallurgy, new networks clearly emerged through which finished metal objects, metallurgical knowledge, and perhaps also raw materials circulated. It has even been suggested that the desire for metal, which took on a particularly important economic and social role in the third and second millennia BCE, led to the emergence of the first era of globalization (Vandkilde 2016).

Numerous hypotheses have been proposed to explain the very earliest experimentations with smelting. Native copper – metallic copper available on the ground surface – was likely the first metal to see experimentation. It could have been shaped into small ornaments or objects, just as stone was shaped into beads and tools through knapping, drilling, and grinding (Roberts et al. 2009). Accidental heating of native copper may have led to the earliest experimentation with annealing; however, given that heat treatment is also deployed in the production of objects in clay, flint, and wood, intentional experimentation may also have been carried out (Radivojević et al. 2010, 2776). Discovering that some stones are also metal ores seems likely to have been a more drawn-out process, but this too has been suggested to originate in pre-existing practices and technologies. Colorful stones, especially green ones, circulated throughout the Neolithic period in the form of axes and ornaments, as well as raw materials to be ground down to make body paint and makeup (Bar-Yosef Mayer and Porat 2008). Malachite, for example, a bright green copper ore, was widely prized around the eastern Mediterranean during the Neolithic, where it was shaped into beads and ground into makeup. Azurite (copper ore) and galena (lead and silver ore) were also in circulation in the later Neolithic Near East, and both were likely ground as colorants, as well as used to make beads. Grinding ore is the first step to successfully smelting it, so accidental or intentional heating of ground colorants may lie at the root of the earliest ore smelting,[2] although this is not supported by archaeometallurgical research in at least one early smelting region (Radivojević *et al.* 2010).

Models of early metallurgy – like the models of early agriculture discussed in Chapter 2 – tend to lean hard on the idea of parsimony: that is, that the successful adoption of a given invention was rare, and that there were likely only few points of origin for early metallurgy. Even as recently as 2009, Roberts *et al.* (2009), for example, were able to publish an elegant and forceful argument supporting a single point of invention for copper smelting in Eurasia. They suggest that copper smelting was invented once in Anatolia and spread outwards from there (Fig. 3.1), based on two factors: the near synchronicity of most of the earliest smelting sites in southwest Asia and southeast Europe, and the presumed complexity of metallurgical processes that, they argue, could have only been successful if carefully controlled and systematically reproduced. While they acknowledge that independent invention occurred – there is

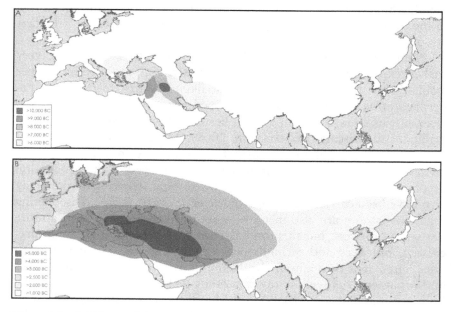

Figure 3.1   (a) The exploitation of copper ores and naturally occurring copper metal; (b) the spread of copper smelting technology.

incontrovertible evidence that gold working, for example, was invented in numerous places – they are skeptical about earlier claims for the independent invention of smelting in Iberia or southeast Europe.

Despite their skepticism, Colin Renfrew's (1967, 1970) model of a hearth of metallurgical innovation in southeast Europe has remained influential, and considerable archaeological evidence has since accumulated that broadly supports his suppositions. Renfrew challenged the dominant parsimonious models by suggesting that new radiocarbon dates indicated the independent invention of metal technology – both copper smelting and gold working – in southeast Europe. He argued against simple diffusionism, pointing out that the tradition of *ex oriente lux* – light from the East – was not well supported by archaeological data and tended to minimize the technological and social complexity found in the Late Neolithic Balkans. Recent archaeometallurgical analysis not only supports Renfrew's independent invention model, but also enriches it by delineating the distinguishing features of the process of smelting in southeast Europe. In a series of articles based on her doctoral research, Miljana Radivojević presents a distinct, and distinctly local, "recipe" for successful copper smelting drawn from evidence dated to fifth-millennium BCE sites of the Vinča culture and based on a technical analysis of the choice and treatment of various copper ores (Radivojević 2015; Radivojević and Rehren 2016; Radivojević *et al.* 2010, 2013). Moreover, she argues that the independent invention of smelting in this region represents only one technological development within a broader polymetallic efflorescence, with smelted copper being produced alongside tin-bronze, gold, lead, and perhaps also silver (Radivojević *et al.* 2013, 1041). Building on work by John Chapman and others (Bailey 2000; Chapman 2007; Gaydarska and Chapman 2008), she argues that the invention of new metallurgical techniques (and associated pyrotechnologies) was spurred by more general social values, including the need for greater forms of personal distinction and wealth as well as an emerging, widespread aesthetic preference for brilliant surfaces and polished textures within the Vinča world of the fifth millennium BCE.

Certainly, in light of this evidence, it must be acknowledged that all of the metal and metalworking traditions identified in prehistoric Eurasia were likely invented multiple times; but we have not yet addressed how prehistoric people might have perceived the emergence of new inventions or the process of invention itself. Radivojević (2015, 322) introduces a broad definition of invention as "a radically new product as much as a recombination of technological components in a novel manner. Alternatively, an invention may involve the application of an existing technology to a new purpose." Yet, the invention she points to in her research – the development of a unique technological recipe for copper smelting – is both temporally and spatially constricted. In her model, the invention is a bounded thing, reproduced more or less correctly by subsequent generations of metalworkers.

Taking a more expansive approach to the temporality of metal making allows us to re-imagine the process she describes as one of constant re-invention. If, following Kienlin (2014, 467), metallurgical knowledge were ritually framed – that is, if the series of steps Radivojević refers to as a recipe were instead a ritual process – then each smelt would have entailed the re-creation of the first successful smelt (i.e., of the invention) with all the uncertainty of outcome and sense of achievement upon success that that implies. Ritual activities often involve an element of temporal manipulation (Gell 1992), including transforming an ancestral "there-then" into a subjective "here-now" (Munn 1992, 113). So, enacting a smelt might well be experienced as flattening time between the metalworker, their teacher, and all the crafters who had previously successfully persuaded stone and fire to make molten copper. In this light, the reproduction of a recipe is also its re-creation, and the embodied experience of the metalworker enacting this process is shaped by a sense of experimentation and hope for – or, perhaps more accurately, faith in – a successful outcome.

Similarly, a wider spatial model of invention might benefit our understanding of how individual prehistoric people would have conceptualized their first interactions with metal and metalworking/-workers. Just as there are many specific hearths of metallurgical experimentation, there are also many parts of the world where metalworking does not seem to have been invented but was adopted in from outside. Both diffusionist and independent invention models of early metallurgy in Eurasia tend to privilege centers of invention, emphasizing their dynamism and regional (or supra-regional) influence while positioning their hinterlands as marginal and passive accepters of foreign innovation. This is most obvious in visual representations where arrows on maps explode outwards from these core regions, pointing to lagging peripheries (e.g., Fig. 5.3) or, in the case of isochrome maps, where solid polygons fade to transparency over marginal areas and regions with limited data (as in Fig. 3.1). The effect of these sorts of illustrations is twofold: first, they imply that marginal regions had little impact on or agency regarding the form or trajectory of newly invented practices and technologies; second, they suggest that the innovation itself remained largely intact and unchanged from its point and moment of invention. They are also largely divorced from prehistoric people's own experiences of invention and innovation, and one is left to wonder whether a person living hundreds or thousands of kilometers from southeastern Europe would understand their first encounter with a copper axe or participation in a smelt as having any connection at all to the time and location of the technology's invention.

Certainly, in central and northern Europe, trajectories of copper adoption suggest a level of local agency and experimentation at odds with the region's apparent position as a distant periphery to more inventive core areas. Metal objects and metallurgy first appear here in the fourth millennium BCE, about a millennium after the Vinča sites discussed above. In the traditional archaeological terminology, this is largely considered to be the Neolithic period but, depending on which part of Europe one looks at, it was simultaneously the early, middle, or late phases of this period, if not the beginnings of the Copper Age (or Chalcolithic) (Roberts and Frieman 2012). In practice, this means that copper objects – if not metallurgy itself – formed part of the Neolithic package that was adopted in northern and western parts of Europe, much the way colorful ores had emerged as significant materials in the Neolithic of the southwestern Levant and southeastern Europe (Roberts and Frieman 2015). This was a complex and heterogeneous period, characterized by the emergence of widespread monumentality and a rich and varied suite of technologies and materials. Pottery was ubiquitous, presumably for use alongside vessels of organic materials, including woven baskets, and leather or rawhide containers. Knapped and ground-stone technologies were sophisticated and widespread, with some objects made from special stone sources circulating hundreds or even thousands of kilometers from their points of origin (Pétrequin et al. 2012). Copper objects circulated into central and western Europe largely from southeastern Europe, particularly along the Mediterranean coasts and via large, navigable rivers (Klassen 2004; Roberts et al. 2009); and the technology was adopted as well, as demonstrated by a few pieces of copper slag excavated at Brixlegg, Austria, and dated to c. 4000 BCE (Höppner et al. 2005). Within a few centuries, we find copious evidence for ore extraction and smelting in the Alpine zone; and copper flat axes, dagger blades, and ornaments were circulating outwards to people living in what is now Atlantic France, the Netherlands, and southern Scandinavia (Roberts and Frieman 2015).

In isolation, the appearance of copper in central and northern Europe certainly looks like a technology derived from elsewhere – the earliest objects known to have been deposited in this region are southeast European in origin (Klassen 2000), and there is no evidence for the use of ore or experimentation with native copper prior to the fourth millennium BCE. Yet, just as to understand the invention of copper smelting it was necessary to explore a suite of related technologies and contextualizing social values – such as the desire for brilliant and flashing materials – so too, if we take a cross-craft and contextual approach to the earliest metal in central and western Europe, a much more dynamic narrative emerges. The rapid adoption of copper-mining and -smelting technology in the Alpine region led almost immediately to a localized system of production specializing in the manufacture of a very restricted range of objects: flat axes, daggers, and ornaments.

Green axes made of special and hard-to-access raw materials – most notable among them jadeitite mined in the Italian Alps and frequently polished to a glassy finish – circulated among Neolithic European communities, from Ireland to Bulgaria and Italy to Scotland, throughout the fifth and fourth millennia BCE (Pétrequin *et al.* 2012). These axes seem to have served both ritual and functional roles, and it appears likely that copper flat axes were sought after at least in part because they drew on the longstanding and deeply symbolic value accorded to their green stone counterparts. By contrast, prior to the introduction of metal there were no daggers in Europe; however, almost synchronous with the earliest metallurgical experimentation, crafters began producing elaborate daggers in flint (Frieman and Eriksen 2015). Although widely considered imitations of metal daggers, instead they seem to emerge from exactly the same sort of technological tradition, and in some places they pre-dated the adoption of copper daggers by generations (Ihuel *et al.* 2015; Steiniger 2015).

In other words, while the first Europeans to smelt copper may have lived in southeast Europe sometime in the fifth millennium BCE, as metallurgy was adopted more broadly it was re-interpreted and re-invented repeatedly to serve and to reflect local needs, social structures, and existing technological practices. New material juxtapositions and social relationships created new contexts of metal use and allowed for locally significant object types and traditions of practice to emerge. Experimentation never ceased, even millennia after the initial moment of invention in southeastern Europe or the Levant. Bray (2012), for example, argues that in Ireland and Britain – where smelting was only adopted in the mid-third millennium BCE – several generations of metalworkers either failed to appreciate or chose to ignore the unique recyclability of liquid copper, choosing to re-cast axes into axes and daggers into daggers, with little to no attempt to treat the metal itself as a combinatory resource. Whilst this might seem rather unsophisticated, these same metallurgists were among the first in western Europe (north of Iberia) to develop a polymetallic tradition, becoming expert goldsmiths and likely inventing a new form of tin-bronze through the alloying of copper and ground cassiterite (Bray and Pollard 2012). At least in the case of metallurgy, the process of invention never really seems to have ended – instead of repeating the process of copper smelting, new metalworkers re-created it, opening the door for constant and further experiments, failures, and inventions.

Invention and innovation

Since Schumpeter's (1934, 1939) germinal work on innovation, there has been a tendency to separate out invention from innovation. Schumpeter positioned invention and innovation as separate elements in the business cycle, with the former having no real impact and the latter being of crucial importance to the development and growth of economic systems. Although more recent research has called his interpretation of economic impacts into question (Parayil 1991), this dichotomy between invention and innovation persists. Yet defining the boundaries between invention and innovation remains elusive. Schumpeter himself struggled to define invention, doing so largely in the negative by suggesting that it does not include incremental and process innovations and is not an economic driver.

Within evolutionary models of technological change, invention has been taken to represent the point of change – the introduction of a new mutation – while innovation encompasses all subsequent elements of communication, experimentation, and adoption: in Erwin and Krakauer's (2004, 1118) words, "invention as origin and fix-ation, and innovation as consequence and success." Roux (2010, 217) suggests that the difference between the two lies in their scale of operation, with invention occurring at the scale of the individual and innovation at the collective or societal level. These formulations all position invention as a fleeting, individualized and momentary event – perhaps part of the reason Schumpeter felt it would have little economic consequence. In doing so, they echo and re-inforce the myth of the solitary genius inventor whose insights are paradigm-shaking epiphanies.

However, although we must recognize invention as a creative and insightful activity, we can also frame it as one that emerges from long histories of working within a particular field, with specific materials, to answer a suite of related questions, and among a community of simi-larly motivated and skilled craftsmen and problem-solvers (MacKenzie and Wajcman 1985, 10). David Gooding (1990), for example, argues against the epiphany model of invention, which he considers to be a product of post hoc, pre-digested, and largely symbolic narratives designed to enhance the reputation of the inventor or the significance of the invention. Based on an extended discussion of the electromag-netic motor, he suggests that the creative work inherent to invention is only possible because of the history of learning, experimentation, and skill-acquisition on the part not only of the inventor, but also of their team and colleagues.

In this, he follows Carlson and Gorman's (Carlson and Gorman 1990; Gorman and Carlson 1990) discussions of the inventors Thomas Edison and Alexander Graham Bell. They approach invention as a cognitive mental process that they divide into three aspects: the mental models used by inventors, which they define as dynamic conceptions of an invention subject to change and alteration as needs and experimentation alter the process; the heuristics or problem-solving strategies utilized by an inventor, and that largely reflect that inventor's own experience, skills, and knowledge; and the mechanical representations available to and developed by inventors, consisting of the specific, working components necessary to build physical models of inventions, such as levers, screws, wedges, wheels, etc. Based on this schema, they suggest that invention does not so much represent a moment of great insight but is best conceived of as a recursive process in which the inventor (and their team) shifts back and forth between abstract conceptualizations of an invention and the tangible prototypes they construct (Fig. 3.2). In other words, invention is neither momentary nor linear, but emerges from a complex process of experimentation, conceptualization, and negotiation.

Invention, then, has time depth, spatial extent, and a network of social and professional relationships. Engineer and historian of technology John Lienhard (2006, 233) has described this model quite poetically as a "filigreed fabric of interwoven invention." I find this metaphor rather apt as textiles are complex, composite objects, frequently multiply

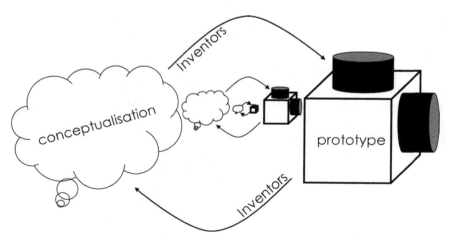

Figure 3.2   Schematic model of a recursive invention process.

authored and regularly cut up, remixed, transformed, and recycled into new forms to suit new needs. Indeed, using textiles as a metaphor for invention also appeals, as textile technology is typically conceived of as women's craft-work. This provides a powerful counterpoint to the mythic and singular heroic inventor, frequently male, who inhabits our technological narratives and appropriates credit for the fruits of collective, collaborative, and complex labor lasting years or generations (Lienhard 2006, 7–8; Wajcman 2000, 453). More than that, from an analytical perspective, a textile is composed of many threads that can be studied individually or collectively to give insight into its creation, history of use, value, and eventual discard. Even if some are lost or damaged, a careful specialist can reconstruct large parts of its fabrication process and finished form.

Thus, as an archaeologist, I find it particularly fruitful to shift our discussion of invention away from moments of insight and singular revolutionary changes and towards a complex process of iteration and experimentation, each phase of which is open to investigation. As we saw to some extent in the discussion of early agriculture in Chapter 2, archaeological narratives of new technologies often tack back and forth between individual instances of very old dates, and regional or even continent-scale maps showing the broad distribution of evidence during a specific, often centuries-long, tranche of time (e.g., Fig. 2.1). This leaves us with the impression of a reasonably uniform, rapid, and linear uptake of new materials and technologies. Such macroscopic models homogenize regional, local, and personal reactions to new inventions, ideas, or practices, not to mention erasing phases of experimentation or re-invention. They normalize the successful diffusion of innovative activities, concomitantly pathologizing resistance and failure – in one example, labeling apparent technological non-change as "cultural stasis" and linking it to eventual cultural extinction (Prentiss and Lenart 2009).

Within the world of early-metal studies, only recently have archaeologists begun to grapple with what it means to reject or resist metal. Silviane Scharl (2016) has documented several pauses or discontinuities in the spread of metal technology and metal adoption. By carefully mapping the dated evidence for metal objects and metallurgy in southeast Europe, she shows that there was a 500-year period (the first half of the fifth millennium BCE) when copper metallurgy was practiced in Transdanubia, but not in neighboring regions to the west. Far from being cut off, ceramic and lithic materials seem to have circulated between these areas, and similar types of enclosures are present across the region. So, presumably, trade and personal relationships existed throughout this period, but metal objects and metallurgy were not part of these networks. We see the same thing happening elsewhere in

Europe: in the early fourth millennium BCE, Pfyn groups in the Alpine foreland were metal-using, but neighboring Michelsberg groups were not; at the same time, copper was being produced in northwest Italy, but neither produced nor used in southeastern France, where in fact the delay in adoption may have lasted until the early third millennium BCE – nearly 1,000 years. So, in these cases, we see a process better characterized as long-term punctuated resistance rather than adoption. Previous attempts at explaining the absence or cessation of metal use have centered around a potential disruption in the supply or availability of the raw materials (Ottaway 1989) or use and depositional practices that were archaeologically invisible (Klassen 2000, 2736).

Both of these explanations are plausible, but both also imply that people must have desired to use and make metal, an assumption I feel we cannot make for past people, who may only rarely have come in contact with singular metallic objects. Instead, Scharl suggests that the punctuated beginnings of metal in different regions reflected a complex and not necessarily identical integration process shaped by demography and social ties. In other words, and returning to our textile metaphor, the need or desire for metallurgy in these peripheral areas might be envisaged as the shuttle guiding the weft of kinship ties (through which metallurgical knowledge appears to have been communicated) as it weaves back and forth through an ever-shifting warp of complex economic, social, and demographic factors (Fig. 3.3).

Imitation, emulation, iteration

A further complication in our discussion of invention is the role of iteration and imitation in the invention process. Popular narratives of invention – the same ones that emphasize singular inventors and eureka moments – also privilege the momentous and momentary invention, regarding its emulation as derivative and less creative. Indeed, philosopher of science René Girard (1990) points to a fundamental conflict between our contemporary notions of innovation and imitation. He argues that this is constructed onto the historical development of technologies and practices, since imitation and innovation are entwined and continuous, only able to be distinguished through later abstractions.

Yet, the idea that invention and imitation are, effectively, antonyms has fed directly into the archaeological discourse. So, for example, we have the case of *skeuomorphism*: meaningful imitation in one material of an object typically made in another.[3] The concept of skeuomorphism was developed to describe a sort of formal or aesthetic "lag," in which new materials and technologies were shaped to look like the older forms they were believed to have replaced (Colley March 1889). Thus, wooden

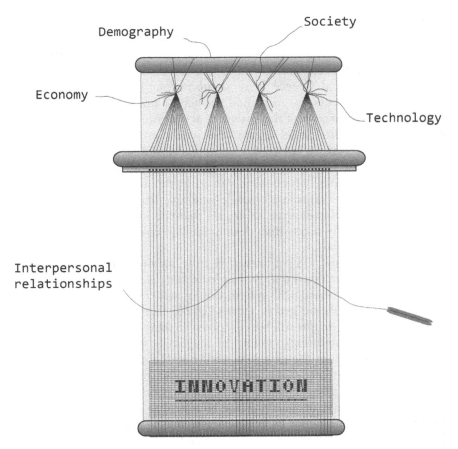

Figure 3.3 The warp and weft of innovation.

forms were retained even when architecture made use of stone, or basketry was imitated in the form and painted patterns on new ceramic vessels. Obviously, this conceptualization of imitation carries with it a value system in which the imitative object is less technologically advanced, less valued, and of secondary importance. In its original formulation, skeuomorphism was thought to represent habits rather than choices, old fashioned tastes rather than creative technologies.

Certainly, this distinction between imitation and original both emerges from and builds on modern conceptions of authenticity, in which the first version of a given thing, practice, or technology is the most creative, the most valued, and the most authentic while later iterations can be anything from aspirational copies to counterfeits or knock-offs intended to deceive and trade on (or engage and play with) the higher value of the original (Chu 2018; Rehn and Vachhani 2006; Thomas 2016).[4] Some archaeological narratives of skeuomorphism still lean on this system of values. So, for example, Michael Vickers and colleagues (Vickers 1986; Vickers and Gill 1994) influentially argued that black ceramics in classical Greece were skeuomorphs of silver vessels: the black glaze represents the tarnished surface of unpolished silver. They suggest that the purpose of this intentional imitation was to increase the value (and, thus, the price) of the otherwise commonplace pottery by creating an association between it and silver. Similarly, Thomas Fenn (2015) has recently made the case that glass was developed intentionally to imitate rare stones, which were valued in the ancient world. Cobalt blues approximated lapis lazuli, while pale blue frits were like turquoise or malachite. In other words, he argues that the inventive process may have been driven by acquisitive desires for less costly and more access-ible prestige materials.

In previous work, I have argued that archaeological discussions of skeuomorphism are often overly simplistic because they do not encom-pass the larger technological system out of which a given object, aes-thetic, form, or tool emerged (Frieman 2012c). The ways that things are crafted affect their final form, but also exist in their own social, material, and technological contexts. Imitation may exist here – in the organ-ization of production, the gestures used to make and craft, or the way materials were conceptualized, for example. Nevertheless, at the same time, morphologically similar things might be produced in wildly diver-gent ways. So, I have argued that the emergence of flint daggers in the fourth and third millennia BCE in western and central Europe was not due to an attempt to copy a newly introduced metal object type, as many previous archaeologists have hypothesized, but was part of a wider cultural shift towards what Vander Linden (2007, 182) has referred to as "stereotyped" artefact assemblages made up of widely recogniz-able objects, if differently manufactured of divergent raw materials. These assemblages included a number of new forms of object, such as V-perforated buttons: archery equipment; various types of ceramic ves-sels; and, indeed, daggers in various materials including copper, bronze, and flint. While superficially similar in appearance and widely adopted, decades of research indicates that these objects manifestly had different uses and meanings in different local contexts across prehistoric Europe.

Vander Linden suggests that stereotyped object forms reflect pro-cesses of categorization related to the governance of the funerary sphere in later prehistory, but my own take on the phenomenon, viewed through the lens of close technological analysis of the flint and metal daggers, is that this desire for similarly shaped objects both emerged from and encouraged the extensive and extending networks of communication and contact that characterized the fourth and third millennia. Over the course of the fourth and third millennia BCE, European people seem to have developed shared ritual architecture, in the form first of earthen long barrows and later megalithic constructions; shared artefact types that became increasingly widespread; and shared technological systems, including the adoption of metalworking, which spread through central and southwestern Europe (and eventually into western and northwest Europe) during these two millennia. Inspired by Kienlin's suggestion that metallurgical knowledge was maintained and mobilized through networks of kin, I suggest that possessing, displaying, and using similarly formed objects to those of distant kin and trading partners (even if, internally to their own respective communities, they used and thought of them quite differently) may have eased some of the difficulties in maintaining personal and economic ties over generations and long distances. In other words, we know those people way over there are our sort of people because they have the same sorts of stuff we do. Moreover, although lithic and metal technologies differ considerably, I have been able to identify a number of similarities in the structure of the technological sequence through which each is crafted, including a preference for special, or specially sourced, raw materials; the use of extremely specialist production techniques; and a high likelihood that at least some of their manufacture phases were conducted away from the wider community, perhaps to limit who had access to the techniques in question. In other words, flint and metal daggers both represent and encompass a shared sphere of technological and social values and practices. In this way, although imitation between both the forms and technologies exists, it is both creative and productive.

In fact, imitation plays an important role in the inventive process. In her germinal monograph on early American technology and invention, Brooke Hindle (1983) has persuasively argued that, in the nineteenth century, emulation played a key role not only in the transmission of production techniques, but also in the invention of new technologies and practices (the role of imitation in teaching and learning is discussed in greater detail in Chapter 5). Novice inventors keen to create their own inventive things would attempt to recreate a master's work and, in so doing, learn its affordances and limitations. In subsequent replications, they would then build in their own adaptations, improvements, and other alterations. This is invention via iteration rather than sudden inspiration, an obvious and complex pattern in our tapestry of inventive acts.

Interestingly, it is within the realm of the digital – a most contemporary and, supposedly, radically new technological sphere – where the power of imitative acts to spur or entangle with invention has been most recently re-discovered. Iteration and creative imitation are part of the native language of digital "Maker" communities. A full half of the designs for 3D models on one digital platform examined by Flath *et al.* (2017) are creative re-combinations – so-called remixes – of older models on the site, rather than novel forms created from scratch or designed largely after inspirations from beyond the platform in question. While the ease of making small tweaks to older designs is likely part of the reason for this proliferation of iterative inventiveness, re-mixing is also a valuable social act, engaging users of the platform with each other through performative re-formulation of the various digital objects they are creating and manipulating.

Stanko (2016, 773) collapses this network of creative engagement with objects, techniques, and fellow creators within the digital Maker sphere into the concept of "knowledge collaboration," whereby the objects themselves mediate the creative relationship between inventors who may never directly interact. The idea that creative acts are inherently socially situated and enfold recombination has also been discussed in the context of artistic production. The argument here is based on a comparison of contemporary and historical art making and makes clear that "combinative creativity" (i.e., remixing) is a core mechanism of artistic creativity and, indeed, that iteration allows artists to refine and experiment with creative works (Sapsed and Tschang 2014, 127–8). Of particular interest to archaeologists, Sapsed and Tschang also note that these combinative acts are not restricted to the content of a given artistic output – the motifs used or perspective employed, for example – but also apply to the tools used to craft it. That is, re-combination can be observed in the technology of art through, for example, the incorporation of old and new knowledge from various differing domains, or the application of tools developed in one area to new contexts or applications.

Indeed, as archaeologists and others have argued, skeuomorphism, iteration, and recombination likely play a further significant role in the inventive process by rendering new inventions familiar enough to be engaged with, valued, and manipulated in their own right. In a short paper, Taylor (1999) discussed the slow process of metal adoption in eastern and central Europe with the aim of disentangling modern ideas about the utility, versatility, and significance of metal from prehistoric ones. In doing so, he found that this long phase of experimentation, re-invention, and patchwork adoption was not purely about metal as a technology, but was also about metal and metal-making as a socially meaningful category. The physical properties of metal – particularly its ability to be recycled into nearly infinite new forms with no trace of the previous shape, function, or manufacture process – were unlike those of any other material in wide use at the time, and Taylor suggested that they were not immediately recognized and valued.

Instead, people had to *envalue* metal, that is, they had to create meaning and value for its unfamiliar physical properties and for the objects that were made from it. More generally, Taylor's concept of envaluation suggests that before a new idea, practice, or technology can be adopted – before it can even be recognized as a thing – there needs to be a social category or role for it to fill. In other words, understanding what makes an invention distinctly new, and what allows it to be desired, adopted, and altered through combinative creative processes, requires us to investigate the pre-existing social and technological contexts in which that innovation might be situated. If there are none, then that will inevitably slow the adoption process, if not de-rail it entirely. Certainly, Stanko (2016, 777) found that, on the digital platform he studied, extremely novel digital objects – ones with few existing referents or little in the way of an established logic – were not comprehended, were rarely re-mixed, and typically were left to languish without engagement.

In fact, in attempting to define an anthropology of technology, Bryan Pfaffenberger made a very similar point. He wrote that:

> to create a new technology is to create not only a new artefact, but also a new world of social relations and myths in which definitions of what "works" and is "successful" are constructed by the same political relations the technology engenders. It could be objected, to be sure, that a technology either "works" or it doesn't, but this objection obscures the mounting evidence that creating a "successful" technology also requires creating and disseminating the very norms that define it as successful. (Pfaffenberger 1988, 249–50)

In other words, the social and technological context of an innovation must be invented alongside the new object or technology because they are as crucial to its adoption as its function, purpose, or the intent of those who introduced it. Evaluation, then, is the invention of meaning, and without meaning no physical invention can be understood.

Certainly, the ubiquity of re-mixing, combinative creativity, and iterative practices in general militates against the mythic narrative of the inventor as singular hero toiling alone and unaided until struck by inspiration. Moreover, the necessity of cultural referents to the very recognition of a new thing or practice as inventive, and the complex networks of communication observed in inventive and creative communities, support the assertion that it may not even be possible for an invention to emerge in such a way. Instead, invention reveals itself to be a messier and considerably more contingent process with a strongly social aspect. Consequently, I argue that invention – both its human and technological elements – is not only accessible to archaeologists working with a fragmented record of the past, but also is considerably more interesting than the longstanding and quite narrow hunt for the earliest instance of a given thing or practice might imply.

Notes

1   Others, chief among them Tobias Kienlin (2008, 2010), would probably disagree strenuously. Certainly, I find Kienlin's kinship-based models of knowledge transmission and specialization compelling and have leaned on them in my own work (Frieman 2012a, c). I will return to Kienlin and discuss his models in more detail in Chapter 5 as part of my discussion of apprenticeship and learning.
2   Indeed, in later periods the same carefully controlled pyrotechnology used to make metal in these regions was also applied to mineral pastes to make beads (Bar-Yosef Mayer et al. 2004). So, the connections between metallurgical technologies and non-metal ornaments should be understood as both longstanding and complex.
3   Skeuomorphism has formed a considerable focus of my previous research. For a more extensive discussion of this phenomenon, its various interpretations, and its application within archaeology, please see Frieman 2010; 2012a, b, c; 2013.
4   Anthropologists have spent decades developing a critical dialogue around the concept of authenticity (e.g., Bruner 1993; Clifford 1989; Fillitz and Saris 2013; Handler 1986; Jones 1993). They emphasize that this concept is itself a cultural construct, largely framed in opposition to an amorphous, non-authentic or innovative Other, and frequently weaponized by the colonial/capitalist gaze. What is authentic might be what is traditional, but it might also be what is deemed to be original or most clearly metonymous of the larger cultural whole. The authentic thing is a powerful tool constructed to do work, be that the assertion of a national identity or the representation of a colonized people. Yet, the authentic is not thick narrative, but a simplified story with a diminished cast of characters. Archaeologists might be wise to take warning from this dialogue: in our search for individual firsts and oldest authentic inventions, each shorn of all context except chronology, we position ourselves not as sociologists or anthropologists of past societies but as tourists looking to experience the least personally challenging narratives, those that are most easily parsed into our own way of life and value systems.

4

Power, influence, and adoption

Why innovate or adopt innovations?

Owing, probably, to the ubiquity of innovation in our wider discourse, it is surprisingly difficult to find archaeological studies of why people innovate or why they adopt innovations when they encounter them. Or, perhaps to turn that around, we archaeologists already have so many common-sense opinions about why people would innovate that we rarely investigate whether these motivations are genuine (or plausible) rather than projected. Certainly, as is the case with all archaeological research questions starting with "Why ...," accessing the specific motivations of people in the past – and especially the deep past – is not straightforward. Yet, as elaborated at length in Chapter 2, it is entirely possible, though laborious, to reconstruct or re-animate the past value systems, social structures, networks of relationships, and communication patterns that would have influenced decisions about adoption and rejection. Even so, adoption is rarely theorized (or even questioned) by archaeologists; it is simply observed.

By contrast, the adoption of innovations is an area that has received considerable attention from business and organizational studies and marketing researchers since, obviously, understanding why a client will or will not adopt a product affects the bottom line. Being driven by the needs of industry and the imperative to develop and distribute ever greater numbers of successful innovations, this research is rather more goal-oriented than analytical, and generally takes for granted that the economic, political, and social contexts of innovation are situated within the present globalized capitalist structures. Nevertheless, this extensive literature provides a number of useful tools for re-evaluating our narratives of adoption. It builds on foundational work, such as Rogers' diffusion of innovations models, to unpack the adoption process embedded within innovation. While Rogers defines categories of adopters – from innovators to laggards – marketing researchers have a vested interest in understanding the demographics of these groups and how they engage with innovations (Singh 2006).

Among the valuable insights from this research is that adoption (like innovation) is not a binary choice, but is a multi-staged process deeply embedded in networks of personal and professional relationships that may or may not lead to successful implementation. Upon introduction to a new product or process, people investigate: they test it out, observe how it operates, make assessments about its complexity and relative advantage, discuss it with colleagues or peers, compare it to other new or pre-existing technologies or processes, and engage with media or expert opinions about it (Shih and Venkatesh 2004). Yet, far from being an easily predictable process, these observations are qualitative and inherently culturally contingent. That is, the trajectory of innovation adoption is dependent both on the wider cultural (and economic and political, etc.) contexts in which the innovation is introduced or encoun-tered and also on the personal and professional networks into which the innovation is introduced and through which it is communicated. For example, in a foundational paper about the adoption of medical technology by hospitals, Meyer and Goes (1988, 900) observe that the adoption process for different pieces of medical equipment will vary not only depending on the needs and constitution of a given hospital, but also because equipment that might benefit one hospital could cause significant problems for another because of different staffing profiles, budgets, community needs, etc. Beyond these practicalities, on an indi-vidual level adoption is also mediated by emotion. Wood and Moreau's (2006) research into the adoption of digital cameras indicates that the emotional responses experienced by people as they learn about and trial new technologies have a statistically significant impact on their decision to adopt and on how they discuss that decision.

Adoption, then, like the broader field of innovation, must be under-stood both at the level of personal action and within wider structural frameworks. For archaeologists, this means that to ask "Why adopt?" we must look not only at the characteristics of the innovation and how it functions, but also at the social context into which it was intro-duced. The necessity of this sort of approach is thrown into relief by the many conflicting narratives of how and why colonized people adopted European goods and practices during the extended period of contact and colonialism in the modern era. On the one hand, this interaction provides archaeologists with a conspicuous and easily accessible case study of innovation adoption, but, of course, on the other it is also deeply complicated by the violence and power imbalances that charac-terized the European hegemonic domination of colonized peoples, not to mention the frequent inability of colonized peoples to present their own experiences of adoption and innovation. Yet it is actually these contradictions – so close to the surface in archaeological studies of contact and colonialism – that make those complex social processes so visible and accessible to us for study.

Innovation adoption at the point of colonial contact

Today people living in Australia use steel axes, not stone ones.[1] Stone axes (and other knapped and ground lithic implements) litter the Australian landscape. Along with a suite of organic objects – woven dilly bags, wooden spears, shields and boomerangs, bone implements, leather and skin cloaks and adornment – they are among the best known and most recognizable archaeological materials in Australia. Evidence for the production and use of ground stone implements is found across the continent, with the flakes and tools from Phase 2 at Majedbebe, dated to *a.* 65,000 years ago, representing the oldest evidence for ground-edge ax production and use in the world (Clarkson *et al.* 2017). Yet, after perhaps 65,000 years of ground-edged axe production, only two centuries after colonisation these tools have been largely abandoned. Certainly, the violence of colonization in Australia – a genocide which included both widespread massacres of Aboriginal people and kidnapping and forced assimilation of Aboriginal children in an effort to erase their languages, cultures, and beliefs – played a role in the loss of traditional knowledge and the diminution of traditional languages, but this does not explain why individual Aboriginal people chose to adopt steel tools after first learning of them. Although its research questions are decades out of date and its framing is borderline racist, Sharp's (1952) well-known discussion of steel axe adoption by Yir-Yoront speaking people from the Cape York peninsula is one of the few scholarly narratives to engage with this process. He notes that stone axes were embedded in complex, long-distance trade and kinship relationships that were controlled by senior men, meaning that younger men and women had to ask permission to use them. When a Christian mission was established in the area, missionaries gave steel axes away, and these were rapidly adopted by those very younger men and women, leading to internal social conflicts and power struggles. In this example, then, steel axes were not technologically superior, but disrupted pre-existing power dynamics allowing for previously subordinate individuals to access more social prestige through the use of a well-known, and culturally significant tool.

Archaeology seems to support the idea that the Aboriginal Australian adoption of European materials was a process marked by both ambiguity and ambivalence. Certainly, Rodney Harrison's (2002a, 2004, 2006) writing on Kimberley points – finely pressure-flaked bifacial spear points, sometimes with serrated edges often made on colorful European bottle glass (Fig. 4.1) – provides an evocative case study. These points are technologically sophisticated and visually striking, and they were a highly sought-after trade good in the nineteenth century. However, they were not produced originally or exclusively for the trade market – with which they only become inextricably entangled after the 1880s. In addition to their use as a marketable item within the white Australian economy, their production drew on and maintained much older technological and cultural practices linked to masculinity, status, and connection to Country (Harrison 2002a, 372). While the use of glass, like the introduction of steel axes, disrupted networks of trade and communication tied to the acquisition of appropriate lithic resources for point production, the continued manufacture of the points

Figure 4.1   A knapped bottle-glass Kimberley point from Bunuba Country.

by Aboriginal men in the Kimberley allowed them both to re-inforce their social identity – severely under attack by the vicissitudes of colonial conquest – and to re-invest with power deep-rooted relationships among people, things, and places. In other words, adopting glass and adapting it to their longstanding knapping traditions was a form of control exerted over the process of colonialism – a means of mitigating or at least shaping its impact. Elsewhere, Harrison (2003, 328) makes the broader point that, in utilizing glass to make traditional Aboriginal forms of artefacts – including, in at least one case, a type of tool that had ceased to be produced several hundred years prior to European settlement – Aboriginal people were subverting the imperial thrust of colonialism, forcing introduced materials to submit to Aboriginal culture, technology, and bodily practice.

Indigenous perspectives on colonialism unmediated by interpreters or white authors rarely survive to the present in written form but can perhaps be reconstructed to a certain extent by archaeologists working alongside and in collaboration with contemporary Aboriginal people. Of particular vibrancy is the wide array of so-called contact rock art recorded in Australia. Globally, invasion and conflict have been widely demonstrated to provoke innovations in art making among people with pre-existing rock-art traditions (e.g., Dowson 1994; Troncoso and Vergara 2013). Within Australia, this body of art includes traditional depictions of Australian fauna, human figures, material culture, and ceremonies alongside paintings of introduced material culture and European people, as well as art made with new techniques and novel materials (Frieman and May 2019; May *et al.* 2013; May *et al.* 2010; May *et al.* 2017; Taçon *et al.* 2012). In Arnhem Land (Northern Territory), Sally K. May, Paul Taçon, and colleagues have spent the last decade recording and studying the contact rock art on Mirrar Country (within Kakadu National Park). At a rock shelter called Djulirri, they have recorded a wide suite of motifs of introduced material culture – from fauna to vehicles to firearms to letters of the alphabet – in a palimpsest with pre-contact art dating back thousands of years. Some of the art records moments of contact well before European colonization; a painting of an Indonesian boat called a *prau* dated to before 1664 CE reflects centuries of intermittent contact with Macassan *trepang* (sea cucumber) fishers (Taçon and May 2013; Taçon *et al.* 2010).

Beyond novel things, Aboriginal painters at Djulirri also adopted new ways of depicting the world around them. They experimented with perspective in paintings of ships (Fig. 4.2) and wagons, and produced a handful of deeply individual portraits, a clear break with traditional ways of painting the human form. This almost exaggerated realism – especially when juxtaposed with nearby paintings of people and bodies

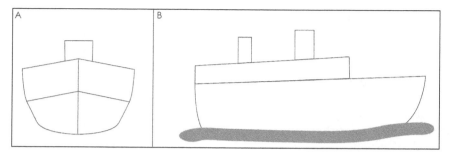

Figure 4.2 Schematic representations of two European ships in perspective based on Aboriginal Australian rock paintings from Djulirri: (a) a front-on view of a steam ship; (b) a steam ship depicted with innovative use of perspective and the waterline illustrated (drawings based on color photographs of these paintings published in Frieman and May 2019, Fig. 6).

that lack it entirely – was restricted to depictions of Europeans, according to information provided by a senior Traditional Owner to Sally K. May (Frieman and May 2019), and may have been portraits of real people known to the painters, or images copied from tobacco tins or other figurative drawings. Yet, despite the flurry of new motifs and art-making practices, at Djulirri, this phase of adoption was fleeting. Traditional motifs and artistic styles were never abandoned, and, after a short phase of experimentation, artists returned to them.

May has argued that rock art was produced in part to educate community members about threats and opportunities (May *et al.* 2013; May *et al.* 2010), so the adoption of new motifs and new styles of depiction likely emerged from this same desire to educate and communicate important information about the new people, things, and ways of life with which Aboriginal people were coming into contact with greater and greater frequency. More than that, depicting new materials using traditional techniques and placing them into art palimpsests alongside and overlain by traditional motifs with deep historic significance is reminiscent of the hypothesized subversion of colonialism enacted when a piece of European glass is forced to submit to Aboriginal knapping practices. The European things, bodies, and practices in this art were objectified, controlled, and contained – forced into co-existence with Aboriginal culture by Aboriginal people. In the case of Djulirri in particular, we know that this process was likely carried out away from European observers. The rock shelter is not near settlements where Aboriginal people and Europeans came into contact, and in fact is semi-isolated from other art-making areas, though near a path used by Mirrar people. This sort of positioning hints at several different possibilities: the contact art at Djulirri may have been segregated or relegated to specific social contexts. It may also be linked to the objectification and assimilation of new technologies and tools; perhaps physically isolating these motifs reflected attempts to control, contain, or limit the impact of Europeans and their material culture to specific spheres. Certainly, based on the long history of art making at Djulirri, which continued after European motifs were abandoned, it was nonetheless regularly encountered, engaged with, and considered by Mirrar people who were themselves negotiating their own identities in the face of colonial invasion.

Beyond Australia, we see similarly complex social and technological networks develop around the adoption of European materials in situations of contact and colonialism, all more or less making clear that the physical properties and functional capacity of new technologies are rarely the instigating element in their adoption. Ehrhardt's (2005, 2013) research into the ways that Illinois people in the North American midcontinent engaged with European copper-base metal objects forms an interesting comparative example because, in this case, access to European materials seems to have pre-dated meaningful contact with Europeans and European culture by decades, if not generations. Copper itself was not new to the region, having a long history as an exotic raw material used primarily in the production of ornaments and associated with cosmological figures and ritualized exchange (Ehrhardt 2009). European-manufactured copper-base metal objects – typically kettles, sewing tools, ornaments, or gun parts – while ultimately acquired from missions or small garrisons, were rapidly assimilated into longstanding Native American networks through which a variety of prestige and exotic objects and materials (locally acquired native copper included) circulated. Moreover, the copper-base objects themselves were rarely used as is, but were more commonly re-worked into a variety of ornaments, including beads, pendants, rings, and bracelets. Adoption, in this case, is almost incidental with the new materials and objects being re-classified into already meaningful categories and re-worked into forms with value to Illinois people. An obvious parallel for this is Harrison's (2002b, 67) discussion of the use of iron by Aboriginal people at Old Lamboo Station (West Australia), in which he notes that iron objects were rarely adopted unaltered, but adapted – either physically or through re-interpretation – by and for Aboriginal people, often in ways that conformed to or reflected pre-contact artefact forms.

Contact between indigenous peoples and foreign colonizers is often (falsely) framed as a moment of rupture, a stark dividing line separating two realities. In the time BC – that is, before contact – indigenous peoples used locally available resources, engaged with each other in traditionally acceptable ways, and used technologies developed locally and with deep and meaningful histories; whilst, in the time AC – after contact – relationships and ways of life disintegrated, local resources were either co-opted by colonizers or abandoned, and newly intro-duced materials and technologies rapidly replaced pre-contact toolkits. This is a narrative of technological and social change that acknow-ledges the profound cultural and physical violence experienced by many indigenous peoples at the hands of incoming colonizers, but it still positions the adoption of foreign practices and technologies almost as an inevitability, a nearly passive process brought about by the combined desire for new or superior technologies and pressure to conform to new, colonized identities (see discussion in Ehrhardt 2005, 12–22). Michael Wilcox (2010), in a searing critique of Jared Diamond's grand narratives of Native American "cultural collapse" and displacement, makes clear that this erasure of Indigenous agency in the face of colonialism is itself an intended and explicit part of that process. Moreover, as discussed in the previous chapter, our narratives of technological change and innovation are themselves profoundly bound up in colonialist discourse. As the examples discussed above demonstrate, there is, in fact, copious evidence that colonized peoples engaged with outside influences, mater-ials, practices, ways of life, and technology in various and complex ways that subverted European norms, created space for indigenous people to comment on these, or contributed to the persistence of indigenous culture and values (or, of course, a mix of all of these) (Bhabha 1994; Flexner 2014; Panich 2013). Adoption within this context is complex, politically and socially fraught, and perhaps might better be understood as incorporation into an existing social and material world rather than an addition to or replacement of it. It reflected more than purely materi-alist evaluations, and likely succeeded as much because of attrition – the violent destruction of indigenous persons, communities, cultures, lan-guages, and histories of technological expertise – as a desire or rational decision making.

Homo economicus the innovator

The archaeological approach to adoption is, deeply unsurprisingly, heavily colored by our interest in the materiality of the past. We focus on what new things do once they are adopted to a far greater extent than we explore how people felt about them or about their adoption. Whilst our sibling social-science disciplines have long moved on from functionalist perspectives in which all elements of human culture – from cutting tools to cars to cosmology – could be understood as fulfilling and responding to a pre-existing social/physical/technological need, many archaeologists still implicitly accept this framing. This approach, of course, reflects the nature of our data. In almost all cases, we lack the ability to interview early adopters; otherwise access their perspectives on newly introduced things or practices; or canvass widely to understand why people choose to adopt new technologies, practices, etc. Archaeology, almost by definition, lacks an emic view. Instead, we must draw our interpretations from things and domains in which they operated. As a consequence, much of our writing around adoption – when we even bother to write about adoption (more below) – is descriptive and inferential. In other words, we work to establish the function (or functions) of a given object, technology, or practice; and, from there, we infer backwards from this point to identify what needs it fulfilled.

So, if we think back to the discussion of early agriculture in Chapter 2, we can easily see what agriculture *does* in early farming communities: it tethers people to place; it provides a (somewhat) predictable and (somewhat) controllable source of food; it potentially (and eventually in some parts of the world) provides greater caloric yields from less territory than hunting and gathering; it serves as a source of wealth. But instead of looking at broader social factors surrounding the phases of adoption and rejection of agriculture, not to mention the opinions of individuals experimenting with cultivation, most archaeologists have assumed that these functions represent *needs*: the need for stable food sources – perhaps due to climate change or increasing demographic pressure; the need for more forms of status or wealth accumulation; the need to exert control over natural spaces. In these explanatory frameworks, a straight line is drawn between needs that are themselves implicitly framed as pre-existing and the functions observed by archaeologists. In the words of famed American archaeologist Lew Binford (1983, 221), "when I am faced with a question, such as why complex systems come into being, my first reaction is to ask what problem people were attempting to solve by new means." Indeed, that an innovation must serve a function at all is such an embedded part of archaeological interpretation that a widespread archaeological joke concerns ambiguous artefacts and practices that must, according to wags, be for ritual purposes if other more prosaic functions are impossible to identify. Thus, the utility of innovations is assumed at every level of interpretation, and these assumptions compound. So, after first interrogating innovations such as agriculture to understand what purpose they served subsequent to adoption (knowing, of course, that they *must* have served a purpose), archaeological models have a tendency to designate this purpose as the core or sole reason for their adoption, effectively erasing any complex and emotionally charged social structures around innovation adoption.

The recent importation of ecological models drawing on niche construction theory into archaeological domestication studies underscores the ways that agency around adoption is replaced by tacitly accepted causal relationships between necessity and function. Niche construction theory (NCT) posits that organisms – in this case humans – are not just affected by their environment, but actively shape it to enhance its suitability for habitation. Applied to the question of how and why domestication began, NCT models build on the available archaeological and environmental data to suggest that the hunter-gatherers who began the process of domesticating plants were engaged in a long, complex, and broad-scale effort at ameliorating their local environments, in part through the specialist exploitation and eventual cultivation of a wide variety of plant species, the less economically useful of which were eventually abandoned (Zeder 2016; Zeder and Smith 2009). Whilst, on the surface level, the push–pull factors assumed by earlier models are absent in an NCT framework, the blending of evolutionary and (modern, capitalist) economic concepts persists, as for example when Smith (2016, 313) suggests that high-value plants were those "with economic utility [that] responded in ways that encouraged and rewarded additional investment of human capital." Even beyond the anachronistic definition of value, this framework implies a certain level of rational and teleological planning to hunter-gatherers' experimentations with cultivation, as it implies they are collectively working from a mental template with common and planned end-points, including a specialization in "economically valuable" cultivars.

This evokes an image that resonates more with modern research-and-development units working to managerially determined future planning proposals than with small-scale societies just developing a new way of relating to the plants in their local environment. Moreover, it leaves little space in our discussion of past people and societies for building complex models of adoption processes that stall, that fail outright, that are successful but only among a segment of the population, or that are successful for a while before being abandoned, since these, by definition, do not function (or perhaps do not function properly, if we are keeping with teleological, needs-based models).

Technological change in particular suffers from this *post hoc ergo propter hoc* "Do–Need" framing. Until recently, our intellectual framework for considering technology and its social context was largely implicit and technologically determinist. Objects and technologies were commonly thought to reflect human action and desire for action passively, rather than emerging from or affecting human relationships in their own right; and technological innovation is assumed to be inevitable (Feenberg 1992; Nye 2006, Chapter 2). Consequently, technological change in the past – as in the present – has generally been treated as an inexorable and progressive increase in functional sophistication and efficiency. For example, a number of studies have been conducted that compare the functionality of stone, copper, bronze, and steel axes as measured by, among other criteria, the speed of task completion (with the task being almost universally chopping down a small tree), caloric output, or time between re-sharpening episodes (Kienlin and Ottaway 1998; Mathieu and Meyer 1997; Saraydar and Shimada 1971). The results of this research largely agree that steel is, in these terms, more efficient than bronze, copper, or stone, and that bronze is more efficient than copper, but that these differences are only apparent when the task's difficulty increases (in these cases, when the tree being chopped down is over about 10–20 cm in diameter).

It is worth noting that we have no evidence whether any of these particular measures of efficiency were valued by any specific group of (let alone all) past people. Efficiency, often presented as an objective measure of intrinsic utility, is itself an interpretation of human action and values based on present-day assumptions (colored strongly by contemporary political economies) about how people conceive of and engage with things. More than that, the *idea of efficiency* is bound up with morality (Winner 1986, 46), and, within the neo-liberal economic models, which have had such influence on how we conceive of ancient people's behavior, it reflects and reinforces the systems that contribute to creating and preserving existing distributions of power and wealth (Barker 1995; Marcuse 1941). Indeed, when archaeological narratives of efficiency are set against ethnographic reports of innovation adoption, the results tend to be contradictory. Townsend (1969), for example, staged an experiment to prove the efficiency of steel axes in a New Guinea community where they had only partially been adopted, even though many members of the village were evidently familiar with their use. Similarly, Sharp (1952) reported little difference in the use or meaning of steel and stone axes by Yir-Yoront-speaking people from the Cape York peninsula, except for the steel axes' ability to disrupt traditional exchange networks. Most tellingly, Bayman (2009, 135) highlights an ethnohistoric report of stone axes remaining the preferred tool of Hawai'ian wood carvers into the early twentieth century, well after the widespread adoption of steel for other activities, because they were thought to be most effective for fine work. Therefore, in assessing why people adopt new things, we must first take care to make clear the assumptions underlying our functional assessments. Not only are the functions of objects, technologies, and societies both socially and temporally contingent; our own economic and technological cultures have thoroughly colored our ability to identify and value these functions for people other than us.

Of course, archaeology is not alone in this sort of technologically determinist approach to adoption. Even in the more human-focused field of sociology, following Wejnert's (2002, 299–302) synthesis, when the characteristics of an innovation itself are considered, the primary variables assessed are the consequences of innovation – its ability to fill a need and its function once adopted – and economic modeling around perceived, assumed, or observed costs and benefits. By contrast, the characteristics of innovations are a major focus for the information technology sphere. Numerous scholars have contributed decades of work to establishing predictive models of adoption behavior grounded in concepts such as ease of use, usefulness, job-fit, and complexity (Davis 1989; Davis *et al.* 1989, 1992; Park *et al.* 2011; Venkatesh *et al.* 2003). Additionally, most of these studies were conducted well past the point of adoption or rejection, so informants' reflections on their adoption rationale were retrospective and digested rather than immediate and unreflected, perhaps leading to more functionalist framing (Venkatesh *et al.* 2003, 437).[2] However, many of these models balance technological criteria with social factors, such as peer pressure or influence; emotional effect or attitude towards use, cultural, or organizational norms; and how the innovation is perceived to contribute to status or personal image. In the widely applied unified theory of acceptance and use of technology (UTAUT) proposed by Venkatesh *et al.* (2003), only two of the four determinants of intention to use/use of technology that were supported by their empirical model – performance expectancy and effort expectancy – related to the function or capacity of the innovation in question. In addition to these, they found that social influence and facilitating conditions, such as environmental factors and fit with wider personal or organizational values, were major contributing factors. Moreover, all four of these determinant criteria were differentially modified by personal attributes, such as age, gender, and experience (Fig. 4.3).

Savvy operators

I recognize that I am walking a bit of a tightrope regarding the agency and knowledge of prehistoric people when I critique the Do–Need framing typically applied to innovation adoption in archaeological narratives. Certainly, as discussed, I feel that our models should not imply that innovations are adopted largely because they fulfill pre-existing needs, including the inferred (but rarely proved) desire for more efficiency; but I also am aware that people in the past – like people today – were skilled at assessing the social, political, and economic norms of their worlds and manipulating these to their own and their kin's or other community's benefit. Whilst *Homo economicus*, the legendary (and entirely mythical) capitalist rational actor, certainly did not exist in the past, that does

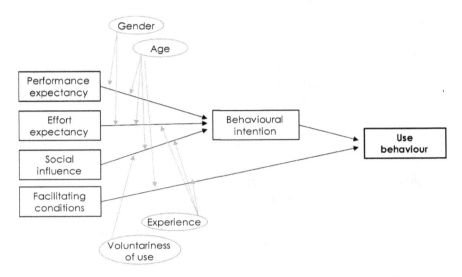

Figure 4.3   The UTAUT model of factors contributing to and shaping innovation adoption and use behavior.

not mean that individuals and groups were not acting in manners that were rational to them – that is, in line with their own social, political, economic, demographic, etc. contexts – when they engaged with new practices, technologies, or ways of life (cf. Dornan 2002).

Although not typically examined within the framework of innovation adoption and diffusion, the emergence of ranked societies and the wide body of literature around the origin of institutionalized inequality and status hierarchies in archaeology proves a fruitful example. The emergence of inequality is a field of longstanding interest to a wide swath of the social sciences, and, among archaeologists, research into this area has followed a number of avenues, with political-economy and evolutionary models proving among the most popular (Ames 2007; Earle 1997; Price and Feinman 2010). North American archaeologist Brian Hayden has elaborated a widely cited, though controversial, model in which "aggrandizers" – ambitious, aggressive, self-interested, almost always male individuals – manipulate their local social structure, economic order, and interpersonal networks to accrue prestige and power to themselves (Hayden 1995b; 2001). Given the right mix of social context, demographic pressure, available technology, and political system, it is hypothesized that they rapidly leverage their local resources into a power base from which they can strike out and accumulate further social, political, material, and economic control (Clark and Blake 1994). That is, they adopt new practices – or at least practices not otherwise chosen by most of their society – with intent, and knowingly manipulate their social network to encourage others to adopt those practices and technologies that will benefit the aggrandizer. This is obviously highly rational – if rather anti-social – behavior and falls under the heading of what Renfrew (1973, 223) terms an *individualizing* form of ranked society.

Hayden and others have suggested that specific newly introduced materials and technologies may have served as "prestige goods" whose acquisition could be limited or controlled by the aggrandizer to maximize his own benefits (Hayden 1998). Moreover, proponents of aggrandizer models often explicitly link successful power accumulation strategies to technological and social structures that make innovation adoption both attractive and frequent (e.g., Clark and Blake 1994, 29). So, for example, Renfrew (1973) links the spread of metallurgical knowledge, metal objects, and new weapon types (daggers in particular) to the shift from *group-oriented* to individualizing chiefdoms in Early Bronze Age northern Europe. Indeed, a number of archaeologists have suggested that the finely knapped flint daggers of the Scandinavian Late Neolithic (2350–1600 BCE) may have been produced by craftsmen controlled by aggrandizing chiefs keen both to display and to enact their power over exchange networks and craftspeople alike (Apel 2001; Earle 2004; Kristiansen 1987).

As discussed in Chapter 3, these objects are traditionally suggested to have been made to imitate and compete with copper and bronze (materials that were not mined in third- and second-millennium BCE Scandinavia, and only infrequently manufactured there up to ca 1600 BCE), as many were likewise made from specially acquired, high-quality raw materials and created in a highly skilled and specialized manufacture process. Yet, recent re-evaluations of flint dagger manufacture and use have highlighted their variable manufacture quality, raw material type and use traces, as well as their extensive links to flint dagger manufacture and use throughout central and western Europe (Frieman 2012a; Frieman and Janz 2018; Olausson 2008, 2017). This research suggests both that the process of their manufacture was neither as controlled nor as specialized as aggrandizer models assume, and that their production likely did not reflect a desire to adopt a prestige good that could compete with copper.

While it is inevitable that people living in Scandinavia at this time were using the tools to hand to manipulate their social structures and wider networks, the linear model of innovation adoption proposed in aggrandizer models (not to mention the central position it gives to individualizing selfishness, as discussed in Chapter 2) is perhaps overly simple and normative to capture the complexity of human decision making reflected throughout the innovation process. This is clearly visible in the repeated pattern of Indigenous ambivalence in the face of outside materials and practices. For example, as my colleague Sally K. May and I have suggested (2019), the innovations in Aboriginal Australian rock art displayed at sites such as Djulirri – particularly the introduction of European objects, vessels, and people – are products of a thoughtful and knowing engagement with new technologies, people, and ways of life, but not ones linked to aggrandizing individuals. Instead, we have proposed that they likely served an educational purpose, building communal knowledge about new people, their animals, society, and material culture so that the wider community could better and more safely engage with these colonizing outsiders.

Traditional adoption narratives (written largely by European colonizers and other cultural outsiders) center values, such as efficiency, that emerge from European knowledge-systems in order to suggest that introduced goods were superior on these grounds. In fact, people with different knowledge systems seem more than happy to continue using traditional tools and materials and re-working introduced goods until they fit into pre-existing material and social categories, as well as to resist or work around the imposition of new practices and ways of life. This too is a rational and knowing approach to the introduction of innovations, but one that responds to a different cultural rationale and emerges from a distinctly different constellation of social relationships, political realities, and technological systems.

Innovation and adoption as social practice

Interpersonal relationships have long been thought to play a key role in the adoption and spread of innovations. Much of the diffusion research, following Rogers' approach, underscores the significance of early adopters, particularly those with stature in the community who are able to encourage others – by argument, coercion, or example – to follow their lead. In a considerable proportion of the literature that leans on this research – sociological, anthropological, and archaeological – these individuals are portrayed or described as community leaders who are more open to innovations because of their higher status and better educations. These elites serve as nodes in complex social networks through which new ideas, technologies, and practices are communicated and by whose influence the wider community comes to experiment with, learn about, and desire innovations. Within archaeology, this conception of the influential elite lies at the heart of political-economy models developed by Earle, Kristiansen, and others (Brumfiel and Earle 1987; Earle and Spriggs 2015; Earle 1987, 1997; Kristiansen and Larsson 2005).

So, for example, Kristiansen and Larsson (2005) suggest that elites in Bronze Age Europe sought and reinforced status through journeys to acquire exotic (and, thus, innovative) technologies, tools, and knowledge. These models build on the work of anthropologist Mary Helms (1988, 1993), who centers the search for and acquisition of exotic (read: innovative or novel) knowledge and its paraphernalia (in the form of foreign goods and objects, for example) within elite strategies to assert and acquire power. This is an approach that positions innovation as disruptive and symbolically freighted rather than frequent and negotiated, and, as such, it conflicts with research emphasizing the ubiquity and mundanity of innovation. Instead, as discussed in Chapter 1, what we see is a research focus on specific types of innovations deemed significant or worthwhile – rare exotica, bronze metalworking, etc. This affects the research results, since these domains are themselves already part of broader narratives about class, gender, wealth, and social significance. The parallel here is between the overtly masculine (and "sexy") product innovation and the frequently overlooked, feminized, and mundane process innovations. Were Bronze Age women sharing knowledge between communities or even traveling great distances to find higher-quality wool or more sophisticated loom weights? There are certainly indications that something like this may have been going on (Frei *et al.* 2015; Frei *et al.* 2017a, b; Sabatini *et al.* 2019); but, as a primary research question, it falls outside the discourse on Bronze Age innovation, which is still overwhelmingly centered on male elites and their shining swords.

A modern example of the power of individuals to act as nodes in human networks, disseminating ideas, practices, technologies, and ways of life can be found in the rise of the social media influencer. "Influencers" have been around since the 1980s as a marketing strategy to create an apparent groundswell of word-of-mouth support for new products or technologies, but the early artificial influencers – typically celebrities or other well-known figures identified and guided by PR and marketing professionals to endorse specific products – have given way in recent years to the emergence of a diverse community of social media influencers, who operate online and tend to emerge from organic social networks, rather than be set before them (Gillin 2007). They use personal branding and large follower counts to attract companies to sponsor their endorsements on Twitter, Instagram, Snapchat, Facebook, YouTube TikTok, or any one of countless other regional or multinational online platforms. Many have gone on to parlay online microcelebrity into book contracts, television programs, or other more mainstream forms of celebrity (Khamis *et al.* 2017). Nevertheless, the majority of social media influencers typically have an out-sized profile in an extremely small niche community, often one with a largely female demographic (e.g., "mommy blogging," "makeup tutorials," "wellness," or "fashion and lifestyle instagrammers"), although men's fashion, gaming, and sports influencers (see, e.g., McLeod 2018) also exist and tend to gather a more masculine audience around them. Some of these influencers are arguably still "elites." For example, the wealthy and connected Kardashian family are also well-known social media influencers who use their reality television shows, personal appearances, pop-culture cameos, and social media accounts to promote Kardashian-branded lifestyle products, from fashion to perfume to mobile phone apps, as well as to offer sponsored endorsements of other commercial products.[3]

Many, however, are not, and their very connection to the communities they come to influence seems to be a large part of their appeal. Following ethnographic work by Abidin (2015; 2016a, b) among lifestyle influencers in Singapore, the relatability and authenticity of these women's public affect (and it *is* an affect: Abidin's research makes clear that these are curated and performed identities) is the foundation of their popularity and of the trust placed in their product recommendations. In Abidin's interviews (e.g., in Abidin 2015), popular Singaporean influencers do not come across as cynical manipulators, but strive repeatedly to emphasize that they feel connected to their community – whom they explicitly term "followers," not "fans," so as to reduce the social distance between them – and their followers too perceive a sort of intimacy in the relationship. While agencies sign and promote individual influencers, they rise to prominence and gain followers (and gain the attention of these agencies) through promoting themselves and their lifestyles – even (or perhaps especially) the trivial and mundane elements – as accessible, believable, authentic, and emulatable (Abidin 2015, 5 [n.p.]).

This is a different sort of social influence from the models of diffusion via educated, worldly elites (typically male). These influencers are perhaps better understood in heterarchical terms as "firsts among equals" or peer-leaders; their celebrity is intense, but its scope is narrow and this is what allows them to be marketed and weaponized by corporate sponsors (Galeotti and Goyal 2009). Individually, they are also, obviously, aggrandizers of the sort discussed above: savvy operators who manipulate their social position to gain power and influence. They use their bodies and their passions to build a following, to whom they then present sponsored content in the form of advertorials and curated aspirational lifestyles (Khamis *et al.* 2017). However, contrary to the male aggrandizer models developed by archaeologists, their currency is social connectedness, rather than economic or ritual control, and they must remain both *within* and *of* their social networks in order to retain power and influence. Rather than "elites," then, we might better understand social media influencers as "social referents," members of a community with greater numbers of connections within their network than their peers, and whose actions and behaviors shape and influence the norms for the whole network (Paluck *et al.* 2016). In practice, this means that, unlike elites who are, to a certain extent, detached from the wider population – by dint of their social status, education, ethnicity, etc. – the majority of social media influencers cannot ever separate themselves from their communities, lest the authenticity and intimacy – the sources of their ongoing and perceived influence – be attenuated.

Despite this heterarchical pattern of influence and leadership, hierarchical models of communication and coercion still dominate our narratives of innovation adoption. So, inventors are championed by early adopters who influence the wider society, and eventually the laggards catch on. Colin Renfrew's model of peer–polity interaction might be one way of envisaging this sort of heterarchical influencing in an archaeological context. He sees social changes, including the emergence of complex socio-political formations, as emerging from "the intermediate scale interactions between local but independent communities ... loosely related but independent, interacting groups" (Renfrew 1986, 7). His model emphasizes that interactions of various sorts – categorized as competition, emulation, adoption, and exchange or redistribution – between adjacent communities of more-or-less equivalent scale and social organization drive innovation in social structure, rather than external influences from outside the local region (Fig. 4.4).[4] Yet this model positions communities as normative and homogeneous, and still assumes decision making is carried out by a small minority who dictate changes to their comrades or subordinates. Important communication happens during mass gatherings – the pan-Hellenic games, religious festivals, battles – rather than on a smaller scale.

As we have seen, the choice to adopt or reject a new practice or technology is complicated and non-linear, so it stands to reason that our social model of innovation also needs further complication. In order to do so, I would suggest we must consider alternative networks that may not respond identically to elites or influencers, no matter their status in dominant political or economic structures. We are, of course, all part of multiple communities – we are members of a family, employees in an organization, practitioners of a skill, players of a game, part of one or more ethnic and religious groups, linked to (or separated from) others by our gender or sexuality. Each of these parts of our identity ties us into different social groups, and often links us to largely separate – perhaps even wholly isolated – clusters of people. Ideas, values, and innovations will spread through these networks in different ways depending on their composition and internal norms.

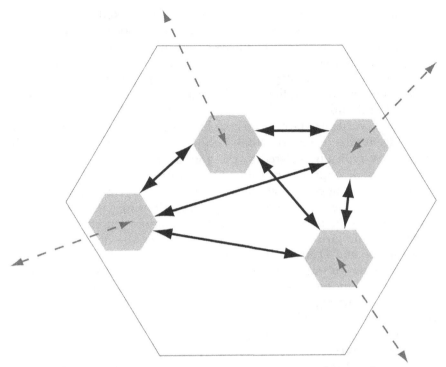

Figure 4.4 Renfrew's model of peer–polity interaction emphasizing the importance of strong ties among closely related, but independent, groups compared to external links.

There are many examples of influential social networks underlying – or even being largely effaced by or submerged within – more dominant structures. By their nature, these submerged networks are composed of non-elites – or at least people who wield little influence in elite-dominated networks – yet they still serve as vectors for the spread of ideas and information. We might look, for example, at the so-called "whisper networks" frequently maintained by minorities or women who use them to spread information covertly to promote personal safety among people like them – typically by identifying sexual harassers, racists, and other predators who are too senior or influential to be formally reported or who are situated within organizations deemed hostile to reporting. Based on anthropological fieldwork in Cairo, Julia Elyachar (2010) has explored the economic and social power of this under-the-radar networking among subordinates (largely women). Through comparing the socializing and kin-building strategies of the wife of a coffeehouse owner with an enterprising madame, she suggests that much of the economic success in the communities where she worked can be attributed to what she terms *phatic labor* – a sort of social infrastructure disparaged by her male informants as "women's talk" (452).

Beyond economic prosperity, these covert, subversive, or simply submerged networks certainly play a role in the adoption of innovations, as we saw in the example of linguistic innovation introduced in Chapter 1. Cross-cultural and historical research suggests that new sounds, intonations, and words are largely invented by and shared within networks of adolescent girls (Cameron 2003; Labov 1990). The innovative speech patterns are typically isolated to these girls' social networks and may even face social opprobrium (often in the form of mockery[5]) from the dominant speaking group, whose elites or social referents are adults and often cis men. The adoption of speech innovations into the mainstream lags a generation, until male and female children raised and taught to speak by the girls who invented them become the dominant group (Tagliamonte and D'Arcy 2009). Furthermore, women's networks are known to play an out-sized role in the uptake and spread of contraceptive or fertility control innovations (Casterline 2001; Kohler 1997), perhaps unsurprisingly, as these largely benefit women's health. Yet, even in the more typically "masculine" domain of agriculture, women's networks have been shown to influence the adoption of innovations, including new farming practices and crop choice (Nielsen 2001). Subedi and Garforth (1996) studied gendered patterns of innovation among Nepalese farmers and found, consistent with other research, that while women were typically less connected and had fewer sources of formal information (e.g., news reports, contact with government initiatives), the heterogeneity of their networks (in that they included both women and men) as well as their willingness and interest in sharing information (and seeds) among themselves contributed to their rapid uptake of farming innovations.

The widely recognized phenomenon of childlore provides another example of efficiency and scope of innovation adoption within a non-dominant or submerged network. Children around the world engage in play and storytelling with each other, sharing skipping or clapping games, rhymes, superstitions, and other forms of folklore that rarely engage with or impact on the adult world (Grider 1980; Sutton-Smith *et al.* 1999). New games and traditions spread rapidly, and co-exist with older ones, some of which persist over tens, if not hundreds, of years (Roud 2011). Some traditions are international – I once found the rhyme of a clapping game I learned as a child in 1980s North Carolina in a museum display about nineteenth-century life in an Australian mining community (neither the Australian nor the Brit who joined me on that visit recognized the rhyme) – whilst others are local. They spread through face-to-face encounters, but are also adapted from television, films, and social media platforms such as YouTube (Bishop 2014). In a widely circulated and deeply moving article in the *Miami New Times*, Lynda Edwards (1997) documented the mythology of homeless children in Florida, whose "secret stories" told of an ongoing war between angels such as The Blue Lady, demons such as La Llorona (or Bloody Mary), and spirits of the recently dead. These stories spread orally from child to child in homeless shelters, and talking about them with adults or even older siblings was frowned upon. Their pantheon echoes older or widespread traditions – The Blue Lady shares many features with a Caribbean Orisha, and invoking Bloody Mary is a rite of passage for children around the world – but their mythology, as documented by Edwards, clearly reflects the violence, disruption, and uncertainty of life as a homeless child, while also offering solace and structure, perhaps contributing to its spread and adoption.

In order to understand adoption, we must move away from linear models rooted in concepts of functionality and efficiency, because these fail to give us insight into the social practices that both underlay and enabled the innovation to be successfully adopted. For example, Papousek (1989) was surprised to see a sudden flurry of innovation in the form of kilns in the 1980s among a small community of Mexican potters with whom he had worked for several years. Although small cylindrical kilns had been standard, larger and square kilns were being introduced. However, none of his informants could give him a clear explanation for why. They suggested square kilns might be less expensive or easier to build, but equally felt they might prove less useful for firing a variety of vessel forms or be more difficult to load. He makes clear that an individual potter's decision to build an innovative kiln type emerges from a complex decision-making process influenced by economic judgments (cost and ease of construction, cost of fuel), family structure (number of family members involved in ceramic production), technological know-how (types and quantity of ceramics to be produced), and his own and his friends' observations of other people's innovative kilns. Yet he argues that this somewhat ragged and incohesive adoption of a new technology represents a collective, if unspoken, effort among the local potting community to free itself economically from intermediaries and middlemen. In other words, adopting the new kiln form represents the final step in several different and possibly contradictory processes of evaluation and observation in which personal relationships are key, but not necessarily determinant or influential.

I want to lean particularly on the element of kin support and interaction because, as discussed above, we too often separate an innovation – particularly a technological one – from the way people engage with it on a day-to-day basis. If we re-conceive of innovations as converging constellations of practice and relationships, then the importance of human relationships looms large, not just as vectors for the transmission of new information, but also as co-producers, emotional support, financial support in case of failure, and trusted companions with whom to compare and discuss one's experiences. Warriner and Moul's (1992) research into Canadian farming practices makes abundantly clear that farmers who farmed with kin – siblings, parents, children, spouses – relied on them for new information and used them as a sounding board while trialing new farming methods. Furthermore, viewing kin support as a crucial element in the adoption process also helps bring into focus the emotional element of adoption that was highlighted in the marketing literature discussed above. This observation is, of course, in line with early diffusion research, which stressed the importance of social factors to the successful adoption of innovations (e.g., Ryan and Gross 1943), but, with the limitations of archaeological data in mind, I want to highlight the importance of these utterly intangible human connections.

Being human, and inextricably connected through social and familial networks to myriad other humans, our reactions to new practices and new technologies are not purely rational: they are filtered through our subjective past experiences, our ephemeral emotional reactions, and the opinions and experiences of those close to us and whose opinions we respect – whether that person is a community leader, a knowledgeable expert, a valued friend, or a tech-savvy member of the younger generation. All of these deeply irrational – and often subconscious – influences color our judgment about more objective values of new practices: that is, how they perform at a given task, their ease of use, or expense to adopt. Yet, their role is often under-appreciated, partly of course because the efficiency discourse holds such sway in our thinking about innovation and technological change, but partly also because our post hoc stories about adoption are necessarily obscured by the processes of closure and stabilization through which opinions coalesce around a particular innovation or solution, rendering it seemingly inevitable in hindsight (see discussion in Chapter 2). These processes collapse the complex social, technological, political, and economic relationships and connections that render the innovation process itself dynamic and fluid so that a messy and irrational process appears, in hindsight, to be obvious and direct.

As archaeologists, we will almost never have a first-person account of why or how a given innovation was adopted. In particularly ancient periods, nearly the whole of past peoples' social networks were submerged, not just those linking non-dominant individuals or groups. Moreover, even for periods in the relatively recent past where considerable documentation exists, such as the period of European colonial contact in Australia, power dynamics between and within the population certainly impact how the narrative is told, whose perspectives are valued, and which opinions are recorded for posterity. Yet, the archaeological perspective can still bring elements of these deeply human, complicated adoption processes to light.

As discussed above, there is ample evidence in both the archaeological and the rock-art record that, in the nineteenth and early twentieth centuries, Aboriginal Australians were engaging with foreign technologies, people, and ideas in complex ways, adopting some and rejecting others in a careful and considered manner. The rock art, in particular, testifies to the complexity of the dialogue happening within and between Aboriginal Australian communities as these newly introduced people and things were examined, tested, compared, and either adopted – usually in a carefully controlled manner – or rejected. This dialogue, and the art that emerged from it, highlights the contradictory and complicated relationship between tradition and innovation that likely played out repeatedly between family members, between friends, between elder and younger community members, and between communities in private spaces and in public actions, such as ceremony. Although those private conversations and individual acts of painting were never documented, the palimpsest of social relationships, public acts, argument, and collaboration can be reconstructed, at least in part, from a careful examination of the rock-art and archaeological records, and in conversation with the artists' descendants. In order to do so, however, we must free ourselves from hindsight and the teleological framework it imposes. Adoption is and was a complex and sometimes convoluted process, but like all human actions it certainly was not linear.

Notes

1 This section draws heavily on research I have been conducting with Dr. Sally K. May (Frieman and May 2019).
2 Using the social construction of technology (SCOT) framework developed by Pinch and Bijker, among others (Bijker 1993; Pinch and Bijker 1987), makes clear that these studies were conducted well past the point of stabilization and closure, meaning the interpretative flexibility that characterizes earlier phases of experiment and adoption had ceased to exist, and the decision around adopting and using a given technology appeared both logical and natural in retrospect.
3 A brief and highly tangential encomium: I once asked a class of students studying British prehistory what made someone "an elite." They were struggling with political-economy models of the Bronze Age and I wanted them to grapple with the complexities of social status beyond "is wealthy" or "is a king." After about a quarter hour of stuttering and near silence from the class, in desperation I asked them if the Kardashians were "elites." The room burst out in shocked

conversation: Kardashian followers were adamantly pro, non-followers equally adamantly con. In the years since, I have used this question over and over and it never ceases to provoke complex and emotionally charged conversation. In developing this section, I felt it was only fair to include them, since they have contributed enormously to my pedagogy. Thanks, Kardashians; I doubt any of you will read this, but your eliteness (or lack thereof) has given a generation of Australian National University archaeology students a really useful way to construct and deconstruct status in the past and present.

4 It is worth noting that this assumption runs counter to longstanding trends in social-network and evolutionary-modeling research that suggest weaker ties are more consequential for social formation and the emergence of innovations than stronger ones (Granovetter 1973; Kuhn 2012; Migliano *et al.* 2020). See further discussion in Chapter 7.

5 A recent and obvious example of this is the extreme over-reaction to creaky voice or vocal fry in English-speaking countries, particularly the USA (Jude 2014). A January 2015 segment on American radio program *This American Life* called "Freedom fries" (www.thisamericanlife.org/545/if-you-dont-have-anything-nice-to-say-say-it-in-all-caps#play [accessed September 15, 2020]) makes achingly clear the virulence of the mockery, as well as its gendered – even misogynistic – flavor.

5

Pass it on

How is the new made known?

The communication of innovations is closely bound up with their adoption: often the mechanism by which one learns about a new thing is also
the impetus for trying it out oneself. So, for example, I spend a lot of
time on social media talking about teaching with other academics; and,
when one of them whom I trust suggests an assessment they developed,
or a new class activity, I am likely to test it. Hence the innovation
spreads. This is a classic case of diffusion via the vector of a trusted early
adopter as described by Rogers (2003). This word-of-mouth approach
to innovation dissemination is common – probably even more so in
the small-scale societies we archaeologists investigate – and reasonably
well understood. As discussed in the previous chapter, trust in and
relatability of the source of an innovation play key roles in the decision
to adopt. So, both the diffusion and communication of innovations are
innately linked to personal relationships and necessarily occur largely
at the local and individual level. Indeed, I think we must also conceive
of this as an active, two- (or more-) way relationship. As anyone who
has ever stood in front of a class has experienced, transmitting new
ideas to others is not an automatic process, but one that requires a
certain amount of gumption and creativity on the part of the knowledge
imparter, as well as at least a minimal amount of interest and attention
from the knowledge impartee.

However, as we have seen in previous chapters, the scale at which
these processes of communication occur and the intangibility of the process itself mean that archaeology finds itself at a disadvantage because
we archaeologists rarely have the data to operate on the personal level.
Consequently, we tend, instead, to focus on phenomena that occur at
the scale of regions or generations. This is why our stories of the communication and spread of innovations frequently gloss over the complexity of transmission processes that are black-boxed in chronological

description (at one date we see this *here*, within fifty years it is also *there*) or subsumed beneath arrows-on-maps models. We tell thin temporal and spatial stories, rather than thick personalized narratives reflecting what must have been contested, negotiated, and not always successful decisions made by many different individuals at different times and places, and relying on very different underlying rationales.

Yet, the two scales are, by necessity, linked. What is widespread diffusion if not a cascading process of single and singular experiences of teaching, learning, and communication? To access and relate these different scales, and to complicate our narrative of learning and adoption, I separate the discussion of innovation dissemination into two parts. First, we explore the complex processes of teaching and learning, and the transfer of culturally significant knowledge among communities and between generations. There is no universal mechanism by which individuals learn a new skill, idea, or practice, but different learning frameworks leave material traces that can, to some extent, be distinguished. Taking a wider perspective on the spread of new ideas, we can also look at models designed to explain their region- and societal-scale diffusion. Although archaeologists regularly observe the communication of innovations on this scale, we rarely contrast different models, tending to anchor our discussions in world-systems approaches, meme theory, or another body of literature without considerable comparative analysis.

To launch our discussion of innovation communication, I start with the spread of Lapita – a ceramic tradition, a cosmological system, and perhaps also a language and people – across the western Pacific. Because the Lapita package spread with people as they colonized uninhabited islands, its dissemination lends itself to linear, holistic models. However, evident regional variation and changes over generations among settled communities complicate the Lapita story.

Ancestral Oceanic migrations and translations

Around 3,500 years ago in the mid-second millennium BCE, a suite of newly combined practices and material cultures developed in the Bismarck Archipelago, east of Papua New Guinea (Fig. 5.1) (Spriggs 1997, 2006). Archaeologists have tended to refer to this emerging group as the Lapita culture or Lapita cultural complex, after the eponymous site in New Caledonia (Kirch 2017; Spriggs 2006), but I will attempt to distinguish between the eponymous Lapita-style pottery and the people who made and used it. The novel social and material repertoire of these people included new ceramic forms with distinctive, dentate-stamped decorative schema (Fig. 5.2); a subsistence economy based on horticulture, arboriculture, and some domestic animals, including

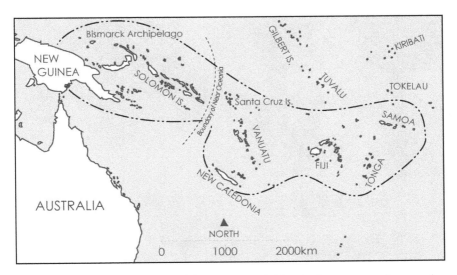

Figure 5.1 Map of the extent of the area occupied by Lapita-pottery-using people.

pigs and chickens; and settlements made up of stilt houses built on beach terraces. Mobility was another distinctive feature of this disparate community: inter-island trade and communication within the Bismarck Archipelago seem to have enhanced social and ritual ties between individuals and lineages.

Within a few generations, the mobility of these Ancestral Pacific Islanders increased, and long-distance, maritime travel became a central feature of these pottery-using agriculturalists' lives. Presumably they traveled in canoes, perhaps similar to later Polynesian outriggers, capacious enough to carry not just multiple individuals, but also seeds, animals, and the basic materials necessary to establish new settlements. Between about 1100 BCE and 900 BCE, they colonized uninhabited eastern islands in Remote Oceania all the way out to Samoa and traveled westwards, with characteristic, dentate-stamped pottery being found around coasts of Papua New Guinea's southeastern peninsula (David *et al.* 2011; McNiven *et al.* 2011). The decorated pottery has been a continuous (and perhaps inordinate; Torrence 2011) topic of research for Pacific archaeologists because of its distinct appearance, making it easy to identify and date Lapita-phase sites; complex decorative motifs, believed to represent a range of socially significant elements, from cosmology to personal or lineage identity; and potential to function as an elite item or trade good – though recent petrographic studies have disproved the latter (Chiu 2012; Leclerc *et al.* 2019b, with references).

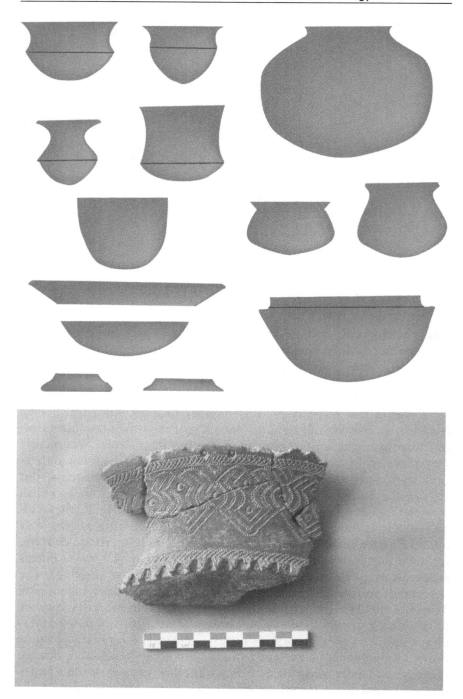

Figure 5.2 (a) Diagram of the range of Lapita pottery forms; (b) sherd of dentate-stamped decorated Lapita pottery from Teouma, Vanuatu.

Until recently, most archaeologists have determined Lapita pottery and ways of life to be quite uniform, especially during the earliest colonization phase of the Lapita dispersal (Golson 1961). However, more evidence of incipient regionalization is appearing as more sites are excavated and more data published (Bedford *et al.* 2019; Sand 2007; Specht *et al.* 2014). This regionalism includes varying ceramic styles, differing technologies (particularly the use of shell tools), divergent histories of chicken- and pig-husbandry, and somewhat different cultivation practices. However, even with this increasingly nuanced research, there remains a clearly shared set of decorative motifs and practices that speaks to a common community or sense of shared identity (Chiu 2012). In fact, a shared characteristic of the well-known pottery is its technological variability: while the forms and decorative schema are quite restricted, there seems to have been no effort to control the actual production of pottery, and early Lapita vessels are made from a variety of clay and temper sources, largely of local origin (Dickinson and Shutler 2000). Within only a couple of generations of colonization, Ancestral Oceanic settlements and practices become marked by increasing regionalization. This is visible in emerging local networks of contact and exchange, such as those through which obsidian circulated (Torrence *et al.* 2018); specific choices in motif used on the decorated Lapita pottery, perhaps linked to house or individual identities (Chiu 2019; Noury 2019; Sand 2007); and local technological traditions, including the increasing uniformity of local potting industries (Summerhayes 2007), suggesting that individual communities developed their own recipes for making ceramics. Bedford and Spriggs (2008) link this regionalization to shifting mobility patterns. Based on their ongoing work in Vanuatu, they suggest that the initial dispersal of people across Remote Oceania was characterized by quite fluid, frequently long-distance mobility between colonizing settlements and outposts, while subsequent generations living in more established settlements saw continued, though less systematic, contact and communication, largely with neighboring communities.

Indeed, this rapid growth in regional practices prefigured the relatively swift abandonment of Lapita pottery and associated practices (and perhaps also identities and political structures). A few generations after the dispersal into Remote Oceania and southern Papua New Guinea, around the mid-to-late first millennium BCE, Lapita pottery ceased to be produced and different sorts of practices began to emerge. Stephanie Cath-Garling (2017) argues strongly against the traditional identification of this process as a decline, devolution, or degradation of the Lapita cultural complex (as in Best 2002). Instead, she suggests that the transition away from Lapita marked a period of re-configured and intensified contact between neighboring and more distant communities in which pig-based status competition, such as is known from the historic period, may have emerged. To her, the abandonment of dentate-stamped Lapita pots in favor of myriad incised, punctuated, and pinched decorative schemata complements shifts in the process of pottery making and represents an escape from the strictures – probably religious as well as related to lineage – that structured and bounded the choices potters could make when decorating Lapita pots.

In the dispersal and regionalization of Lapita and post-Lapita technologies (and the people who used them), we can explore both the spatial spread of innovations between communities – such as the new post-Lapita pottery decorations that characterize the transition period of the late first millennium BCE – as well as processes of teaching and learning through which specific practices and ways of life were (or were not) faithfully passed from one generation to the next over time. A first hurdle, however, is to identify people and social ties in an archaeological literature dominated by categorical thinking that typically identifies dentate-stamped pots with Austronesian speakers presumed to share a specific genetic heritage, and thence with bounded populations (Kirch 2017, 94; cf. the critique by Terrell *et al.* 1997). An example of this is the widely cited *integration/innovation/intrusion* ("Triple I") model proposed by Roger Green (1991), that sees the various elements that make up the Lapita cultural complex as discrete and either pre-existing (belonging to Austronesian-speaking immigrants from Island Southeast Asia or to the local Papuan-speaking people of the Bismarck archipelago) or emerging from the encounter between these two presumed-bounded population groups. This sort of framework misses out the complex interplay between thing and process as well as the messy social relationships that shape bigger systems, such as agriculture, as discussed in Chapters 2 and 3. Moreover, it does not allow for the sorts of patchwork adoption or re-interpretation presented in Chapter 4. Indeed, as Specht *et al.* (2014) make clear, there was no unified Lapita package, and there seem to be clearly different expressions and levels of adoption of Lapita material depending on where one looks in Near or Remote Oceania or Papua New Guinea.

Thus, even in what might seem a classic example of innovations spread through migration, the pattern of innovation dissemination is both patchy and varied, rather than uniform and comprehensive. A similar pattern is visible when looking at the spread of post-Lapita practices and materials. Cath-Garling (2017, Chapter 2) argues that the more or less synchronous "pulse" of new practices for decorating ceramics (e.g., incision, applied relief, fingernail imprinting) suggests that a shared suite of practices and ideas were adopted rapidly throughout much of island Melanesia after around 2,350 cal BP. She suggests that this process of innovation adoption – or social transformation, in her words – occurred at several scales and through various mechanisms (Cath-Garling 2017, Chapter 11). At the macro-scale, the spread of these new motifs and vessel forms – as well as an emergent rock-art tradition – follows a broad trend of human mobility from west to east, from the Bismarck archipelago to as far east as New Caledonia. Within this macro-zone, she identifies meso-scale spheres of more intense interaction among smaller groups of communities visible through shared ceramic vessel forms and decorative choices, as well as the different circulation routes of obsidian and red ocher. While these interaction networks overlap and intersect in places, they do not form bounded or wholly separate regions or obvious inter-island communities. Finally, she describes a micro-sphere of highly localized differences in ceramic style and manufacture technique, e.g., through the choice of tempers. In other words, although she takes pains to say that she does not associate this innovative pulse with the migration of a single ethnic group, she broadly sees increasing, and increasingly intense, mobility and interaction as the mechanism behind the spread of these post-Lapita innovations. She hints, though does not quite state outright, that this intensification of mobility may have been a knock-on effect from changes to Indonesian exchange relationships linked to the adoption of metal.

Unexplored in this discussion is the complex question of kinship. If Lapita represented a cosmological tradition as much as a shared way of life, then its abandonment likely had other less tangible effects that may well have included changing patterns of marriage, adoption, or kinship recognition – all of which could have had significant impacts on the patterns by which people traveled. Similarly, we might see the abandonment of Lapita motifs as the disintegration or abandonment of the houses or lineage groups that they are thought to represent (Chiu 2012), a state of affairs that would allow not only for an influx of new ceramic styles and manufacture processes, but also for the construction of new relationships between individuals and communities. Perhaps the increase in pig remains that Cath-Garling identified reflected a process of relationship building and consolidation between previously unrelated or unconnected communities. This sort of small-scale interaction could conceivably have accounted both for the movement of individuals – marriage partners, adoptees, or other new kin and trading partners – reflected in micro-scale innovations, as well as the emergent networks of communication that underlay the meso-scale mobility of practices and raw materials. I would further suggest that instead of as a "pulse," we might more productively conceptualize this pattern of messy, overlapping, and non-contiguously mobile innovations as more like a rainforest canopy: although, from above, it appears solid and homogeneous, from below and within, its heterogeneous and complex composition is more obvious.

Less studied than the spread of Lapita and post-Lapita materials is their reproduction within communities over time. In this case, since the literature is vast and somewhat fragmented, I will focus only on the pottery. A number of archaeologists have studied the production of pottery as a means of investigating intergenerational knowledge transfer, the creation and maintenance of technological traditions, and the spread of innovations (e.g., Brown 1989; Crown and Wills 1995; Crown 1994, 2001, 2002; Neff 1996; Roux 2003; Shennan and Wilkinson 2001; Stark 1991). One thing that becomes clear in this literature is that there is no universal process through which one learns or teaches pottery making, and different transmission traditions fare equally well at teaching complex technological practices. Historically, anthropologically-oriented archaeologists have been prone to think about potters and potting traditions as conservative (Neff 1992, 163–4 with references), though Patricia Crown's (2001) work in the American southwest highlights the extent to which different cultural practices of apprenticeship or craft teaching affect or permit innovation in pottery making. Based on experimental work and a comparison of ethnographic data, she suggests that formal instruction tends to reduce the variation between teacher and student, while trial-and-error learning leads to more creative and variable results. These two processes are not mutually exclusive and can, in fact, coincide.

It is widely accepted that there was little economic or social control on pottery production in ancient Pacific communities (Leclerc *et al.* 2019a, 14 with references). Indeed, particularly in early Lapita contexts, the ceramic repertoire seems to be characterized by variety in form as well as paste – raw-material choice appears opportunistic, rather than considered (Summerhayes 2001, 2007). We might see this variation as resulting from a rather undirected process of knowledge transmission, perhaps structured like the one Crown hypothesizes for Mimbres potting communities. Following this model, unskilled potters (probably children) would have been provided raw materials and space to experiment alongside more accomplished potters but without guidance. Within this sort of unstructured environment, the enormous variation that characterizes Lapita wares seems likely to result. That the ceramics became more standardized over time likely reflects a growing knowledge of local resources as settlements become more established, and perhaps also subtle shifts in the practice of teaching pottery manufacture.

By contrast, the dentate-stamped motifs applied to the exteriors of some pots are more cohesive, and more control appears to characterize their transmission (Sand 2007). Certainly, if the decoration played a role in identifying lineage or local identities, uniformity and consistent transmission may have been highly valued. Since the decorated pots were likely deployed during public ceremonies and, at least at Teouma in Vanuatu, seem to have been used for presenting or serving a single type of food (Leclerc *et al.* 2019b; Leclerc *et al.* 2018), a ritual element may have also been part of their process. For example, not only might the learning process for ceramic decoration (as opposed to ceramic manufacture) have been carefully controlled, it may have been secret, or constrained to specific individuals, lineages, etc. In other words, the group of people involved in ceramic production might shrink for certain secret or specialized processes. Crown (2016, 72) specifically highlights the production of pottery for ritual usages as one context in which "secrecy and concepts of ownership of knowledge are most likely to emerge." In other words, although Pacific archaeologists tend to collapse pottery decoration and pottery manufacture together into a single technological process, a transmission-focused approach not only disaggregates them, but also suggests that they emerged from separate practices and value systems, as well as perhaps even being carried out by different or more restricted groups of craftspeople. From this perspective, the disappearance of dentate-stamped Lapita motifs might reflect the loss of secret knowledge – perhaps because of its restriction to a very few individuals – or changing cosmological systems that rendered the accurate transmission of these motifs less important than for previous generations. In other words, we can assume that knowledge of ceramic technology was widespread and uncontrolled, but that decoration with dentate-stamped motifs was restricted to a specific, small group of individuals who transmitted it selectively. If these people died without teaching an appropriate apprentice, or the ritual system that made their ownership of these motifs significant lost influence, then maintaining that careful process of teaching and transmission would become difficult and perhaps unnecessary. Future research into the *chaînes opératoires* of Lapita ceramic production and decoration – as well as other, contemporary technological domains – might help clarify and refine our understanding of these processes.

Knowledge transfer, teaching, and learning

How knowledge moves from one individual to the next, one generation to the next, or one group of people to another is a topic of considerable interest to scholars in a variety of disciplines. Although they are intrinsically linked processes, for the purposes of discussion I distinguish between the communication of innovations over time (sometimes called "vertical" knowledge transfer) and their dispersal in space (or "horizontal" knowledge transfer). Biologists study processes of teaching and learning between animals to understand non-human communication and explore instinctual versus learned knowledge (Caro and Hauser 1992). Studies of teaching and knowledge transfer among primates and monkeys give us insight into how our own hominin ancestors may have behaved, and in what ways early human technologies were developed, practiced, and learned (e.g., Derricourt 2018). Neurologists and cognitive scientists have developed a rich body of literature on the physiology of learning, which anthropologists of education such as Jean Lave have used in developing more social models of the teaching and learning process (Lave 2019; Lave and Rogoff 1984). Indeed, anthropologies of apprenticeship and social learning are well established and supported by a number of ethnographic studies from communities around the world (Coy 1989; Lave 2011). Amongst these, Lave and Wenger's (1991; Wenger 1998) ethnographically informed concepts of *situated learning* and *communities of practice* are particularly well known and have been widely applied in professional management contexts (Wenger *et al.* 2002). Within archaeology, much of the research into knowledge transmission or the communication of innovations and practices between individuals and generations has taken the form either of apprenticeship studies or of more complex evolutionary modeling.

Apprenticeship, then, seems like a logical place to start this discussion. In the contemporary literature, apprenticeship is typically contrasted with schooling. While the latter conceives of knowledge, usually written, as discrete bundles of information to be assimilated and deployed, the former, it is proposed, is built around a more holistic concept of learning and practice. Apprentices do not just learn by rote, they learn by doing: knowledge gained through an apprenticeship is necessarily embodied, so the apprentice is not just trained, but encultured and enskilled. That is, their education is about not just learning a skillset (the techniques of the craft), but learning a social role (the job of the crafter), a series of gut reactions appropriate to it, and the formal and informal practices that shape one's ability to act in that role (Marchand 2008). Gísli Pálsson (1994) explores his own experiences of learning deep-sea fishing from Icelandic mariners and suggests that becoming skillful is like getting one's sea legs: a process in which one adapts to and immerses oneself in a new natural and social environment through practice and active engagement.

Some archaeologists, particularly those working within the framework of francophone technology studies, have taken a similar approach. Their emphasis is more on the relationship between knowledge (*connaissance*), conceptual information about production steps or processes that can be taught orally or by example, and know-how (*savoir-faire*), the embodied practices, gestures, reflexes, and muscle memory that underlie and shape technique and that can only be learned through experience (Pelegrin 1990, 118). It is widely assumed that kin relations structured most educational activities in small-scale, ancient societies, and that many of the learning and teaching activities in which children and other novices would have participated were largely informal. Most children would have learned to work wood, knap flint, or make pots through imitation of adult behaviors in play with each other as well as alongside parents, older siblings, and other community members (Arnold 2012; Arthur 2018; Baxter 2005; Högberg 2008). Although observation and imitation were likely of considerable significance to the learning process in these sorts of small-scale contexts, as Crown (2002) reminds us (and as discussed in Chapter 3), there is no reason to expect that this would have impacted on the ability of crafters or novices to invent or innovate. Indeed, following Tim Ingold (2000, 37), these are themselves creative acts, as "to observe is actively to attend to the movements of others; to imitate is to align that attention to the movement of one's own practical orientation towards the environment."

More complex technologies or technological systems are typically expected to have necessitated more structured training regimes. Metalworking, requiring both rare raw materials and specialized practices, is the classic example of a technology that most archaeologists believe would have allowed for or even made imperative the emergence of specialist master crafters who would have trained small groups of apprentices (Ottaway 2001). However, Kienlin (2008) has argued that this specialist knowledge could have been preserved and transmitted by kin within delimited lineage groups, and recent work on the variability of Bronze Age smithing skill supports his suggestion that there was no central control over metallurgy at this period (Kuijpers 2018). Kathryn Arthur's (2018, Chapter 4) observations of teaching and learning among Gamo leatherworkers in Ethiopia supports this model. The leatherworkers she worked with passed a variety of complex, cross-craft knowledge through male lineages, making use of a specially developed argot of technical terminology to protect the secrets of their craft (131–2).

There is perhaps a parallel to be found here between the various skill levels of Bronze Age smiths and the difference in technique and skill that Crown (2016) believes were used to make redware and whiteware ceramics among prehistoric Pueblo potters in the American southwest. Crown suggests that the same communities of potters were responsible for both the very specialized redware and the much more variable whiteware. However, more control was exerted over the production recipe of the former, as its production was limited to initiated members of a specific religious group, while the latter was produced openly. Similar principles are at play in the production of art in Aboriginal Australian communities. May (2008, 182) describes initiated elders controlling the production of art in a contemporary art-making community in Arnhem Land, Northern Territory by making sure that young men wanting to become artists understood the cultural protocols that govern the ownership of motifs, techniques, and stories by different clans, as well as monitoring their output to ensure they complied with these. Thus, specialization, or achieving what Maikel Kuijpers (2018, 563) refers to as "virtuoso" skill, could well reflect age, experience, and commitment to specific non-technological principles (initiation, membership in ritual groups) rather than formal systems of apprenticeship or economic structures that recognized master and novice crafters.

While often conceived of as a hierarchical social formation, apprenticeships can take many forms. Stephanie Bunn (1999), for example, draws out a detailed discussion of how Kyrgyz nomads develop specialist skills through informal systems of learning and teaching. Within this community, most learning is familial. Children emulate adult practices in play from an early age, and their moral and social upbringing are synonymous with their wider education and skills training. In this context, craft specialists are not apprenticed as such, but learn their skills through reciprocal family relationships. In Bunn's (1999, 84) estimation, "it is difficult to draw the boundary between the child and the apprentice, or between upbringing and teaching." Skills are learned for the enjoyment of it, to develop a "birth talent," and as part of the natural process of interchange and engagement between kin and community. Formal apprenticeships may emphasize the social difference between master and apprentice, but even here more complex relations may emerge. John Singleton (1989, 22) notes, for example, that even within the strictly hierarchical context of a Japanese master potter's workshop, the apprentice is like a new bride and their apprenticeship like an adoption into a craft-kinship with their fellow practitioners. The implication here is that the establishment of apprenticeships is not purely a matter of economics or knowledge transmission, but one in which social relationships are created, developed, and maintained through shared practice and knowledge.

Indeed, as discussed in the previous section, systems of teaching and learning vary immensely from people to people, craft to craft, and period to period (not to mention class to class, gender to gender, etc.). I highlight the variability and social complexity of apprenticeship in particular because this militates against the rather played-out concept that these are somehow innately conservative systems that do not allow for invention or creative adaptation to new social, technological, or economic contexts (e.g., Smith 2019). In a germinal paper exploring the role of guilds and apprenticeship in medieval and post-medieval Europe, S. R. Epstein

(1998) argued that, contrary to the received wisdom dating all the way back to Adam Smith's *Wealth of Nations*, guilds and guild-sanctioned apprenticeships fostered innovation and creative problem solving. He argued that regulations restricting the sharing of guild secrets and adoption of outside practices could not have been stringently enforced, and, moreover, that there is a clear and documented history of obvious craft innovation within the guild system that belies this assumption. Epstein suggests instead that tensions between innovation and conservatism did not reflect the structure of training or organization of labor within guild-backed apprenticeship systems, but instead were more likely to be products of relations between guilds and local government, or between richer and poorer crafters. Indeed, he goes so far as to argue that these apprenticeship systems actually fostered craft innovation, since they tended to be established in proximity, creating geographical clusters that translated to localized centers of expertise, specialization, and adequate raw materials and machines for experimentation to be financially viable.

This approach prefigures and echoes Lave and Wenger's (1991; Wenger 1998) model of learning and teaching as ongoing, collective activities that emerge naturally from our engagements with the people around us in what they term *communities of practice*. Following Wenger (1998), communities of practice are ubiquitous and overlapping networks of people through which information flows. They are heterogenous and dynamic, with fluid membership, and are further complexly entangled into broad constellations of related groups with shared histories, trajectories, aims, and/or membership. This model positions mobility and flux as normative, allowing us to investigate how social relations, specific histories of practice, and external pressures combine to favor innovation or continuity. Wenger (1998, 93–4) views continuity itself as a process of active re-invention in which people adapt histories of practice to changing situations, contexts, and networks of relations. Following Brown and Duguid (1991), communities of practice provide a key mechanism by which innovations are introduced into more rigid or static organizational structures, being informal and fluid networks of shared experience and ongoing learning and adaption. They not only subvert centralized power by allowing alternative venues for expertise to develop, but also foster information flows not mediated through organizational bureaucracy (Cox 2005, 535).

This model has been widely applied by archaeologists seeking to understand how learning and knowledge transmission operate and relate to technological change and continuity (papers in Arnold 2012; Roddick and Stahl 2016). Barbara Mills (2018), for example, draws heavily on Wenger's conception of *boundary objects* in her discussion of ceramic innovation and intermarriage in the pre-Hispanic American southwest. A boundary object, as first described by Susan Star (1989, 2010; Star and Griesemer 1989) is a thing (tangible or intangible) that can bridge divergent communities, even as it may hold different meanings or relate to different social processes within each of them. In my own work on European Neolithic and Bronze Age flint daggers, I have suggested that the dagger form (i.e., a pointed, double-edged blade with a handle) itself was a boundary object as, even though so-called daggers have been shown to play different functional roles at different times and places, the form of a double-edged blade seems nearly always to be bound up with ideas of specialization and innovative technologies (Frieman 2012a). Mills builds on Wenger's approach, in which boundary objects are those things, practices, activities, or people that participate in multiple networks or communities of practice. She identifies ceramic serving bowls as boundary objects in Ancestral Pueblo communities, as they were recognizable in form and function but could be produced from different technical recipes and decorated in ways that reflected the different backgrounds of the (likely mobile) potters who produced them. She connects this observation to the wider diffusion of innovations between and among Ancestral Pueblo people and their neighbors through intermarriage and female exogamy, and notes that these sorts of boundary objects would have facilitated the transmission of new information between different communities' potting traditions by easing the transition of individual potters when they left home.

I find the fluidity of these social models considerably more compelling than the more categorical evolutionary models of knowledge transmission that also proliferate in archaeology. Dual inheritance theory posits separate mechanisms for variation over time through knowledge transmission. Social learning – vertical, horizontal, or oblique – allows cultural information to be transmitted from one individual to another and creates a mechanism for traditions to persist and spread over time and space (Boyd and Richerson 1985; Shennan 1989; Shennan and Steele 1999). Richard Dawkins' idea of the meme – a cultural equivalent to the gene – played an important role in evolutionary thinking about cultural transmission (Dawkins 1976; Shennan 2002a, 46–48; 2002b). In the 1990s, Ben Cullen (1993a, 1995, 1996a) proposed a modification of the meme idea that he termed *cultural virus theory*. This model of innovation was based on an analogy between the transcription of viral RNA within infected cells and the transmission of knowledge between individuals and generations. Unfortunately, his sudden death meant that Cullen had little time to respond to early critiques of his model (e.g., Ingold commenting within Cullen 1993b, 195), and this model was never fully realized or applied.

More recent evolutionary approaches tend to nuance or even reject the idea of the meme as overly restrictive and proscriptive. Peter Jordan (2015, 13), for example, suggests that memetic transfer should be expanded to encompass explicitly "skills, beliefs, values, and attitudes" that are not necessarily faithfully transferred between individuals. His model, however, still assumes that a more-or-less discrete and bounded package of information can be transmitted with fidelity from one generation to the next. Indeed, for evolutionary models to work, high-fidelity transmission is necessary. This is a key element in cultural cumulation and ratcheting – that is, the accumulation of learned skills over generations (Lewis and Laland 2012; Tennie *et al.* 2009). The tension in this model lies in the dueling assumptions (a) that most social learning is effectively replicative, and (b) that ratcheting and variation are inherent to cultural evolution – that is, that change is normative. The conflict here lies in a push–pull between conformity and innovativeness that evolutionary theorists solve by identifying a variety of pathways by which innovation might occur (Table 5.1). Innovation enters this system through discrete mechanisms, including stochastic change (drift) and random mutation, but also intentional, biased transmission in which an individual adapts or alters the cultural information they have received

Table 5.1 Shennan's typology of modes of cultural transmission (Shennan [2002a], 50, Table 4).

	Vertical (parent → child)	Horizontal/ contagious	One → many	Concerted (many → one)
Transmitter	Parent(s)	Unrelated	Teacher/ leader/ media	Older members of a social group
Transmittee	Child	Unrelated	Pupils/ citizens/ audience	Younger members of a social group
Acceptance of innovation	Intermediate difficulty	Easy	Easy	Very difficult
Variations between individuals within population	High	Can be high	Low	Lowest
Variation between groups	High	Can be high	Can be high	Smallest
Cultural evolution	Slow	Can be rapid	Most rapid	Most conservative

prior to (or in the process of) transmitting it (Richerson and Boyd 2005; Tehrani and Riede 2008; Whiten 2011).

In fact, this contradiction can easily be solved by stepping away from evolutionary analogies and returning to Lave's (1996, 2009) singularly social observations. Her insistence that we decenter *teaching* and *transmission* in favor of learning and learners emphasizes that learning itself is transformative. Situated learning comprises considerably more than the transmission of discrete bodies of knowledge or packages of information; it socializes the learner and the teacher both, by engaging them in a mutually constituted and dynamic relationship. Moreover, as learning occurs in dynamic communities, those communities, as well as the individual relationships that criss-cross them, are also transformed in the learning process and transform the learned material in turn. Put baldly, not only are we not machines that replicate perfectly until a fault or new program is introduced, knowledge transfer itself transforms knowledge, birthing invention and making space for creativity (Wilkins 2018). Indeed, as discussed in Chapter 3, every imitation re-invents and re-mixes its prototype; emulation is not mechanical but transmutational.

Spreading innovations

For much of the later twentieth century, innovation studies and diffusion studies were inextricably bound up with each other (Hall 2006). Although, today, we tend to think of Rogers' *Diffusion of Innovations* as a monograph about innovation, in fact he couches his approach in this classic book as an investigation into diffusion ("the process by which an innovation is communicated;" Rogers 2003 11) and spends a considerable portion of the text detailing social networks and channels through which innovations flow laterally between individuals and communities. Social geographer Torsten Hägerstrand (1966, 27) took this maxim one step further, declaring that the diffusion of innovations was not simply a process of communication, but was actually the product of interpersonal and intercommunal communication. To develop this concept, he modeled the spread of specific technological innovations (telephones, tuberculosis controls, tractors, automobile ownership, etc.) over time and space in an area of southern Östergötland in Sweden through conventional distribution maps and isarithmic maps (typically called heat maps) that plot variations in the density of mapped phenomena (Hägerstrand 1967). In developing his models of innovation diffusion, Hägerstrand attempted to bring together localized data about population movement and migration, land ownership, property type (farm or non-farm), and networks of communication. His work, like

Rogers', emphasizes the importance of influential community members in the diffusion of innovations. People living in proximity to someone who has adopted an innovation are more likely to adopt it themselves; this "neighborhood effect" is due to the transfer of what he calls "private information" – that is, personal experiences and opinions about the innovation, its value, and its effectiveness. This approach proved influential, with numerous subsequent studies in geography, economics, and sociology emphasizing that propinquity eases communication and speeds the diffusion of innovations (Wejnert 2002, 311–12 with references).

Following this logic, denser areas with less friction in their communication networks should be predisposed to disseminate innovations with greater rapidity than less populated and less connected regions. Furthermore, models of path dependence suggest that the networks that facilitate innovation become cumulative over time, expediting knowledge transfer further. This assumption is frequently brought to bear to explain the uneven geography of innovation: that is, why some places – typically dense, urban centers – seem to foster greater inventiveness or adopt more innovations than others (Martin and Simmie 2008). The observation that innovations cluster (both in time and space) finds its origin with Schumpeter (1939, 100–1) who attributes the uneven spatial distribution of innovative behavior to the regional patterns in which industries develop with regard to each other and to their wider economic and environmental surroundings.

In more recent years, attention has focused on describing and understanding how such clusters – frequently termed *regional innovation systems* (RISs) – develop, operate, and promote innovation (Asheim and Gertler 2009). A core tenet of RIS research is that propinquity is a fundamental feature of innovation because so much of the information flow that underpins innovative behavior is tacit: that is, informal, not codified, and difficult to exchange over long distances. Moreover, RIS models posit that knowledge itself can be "sticky" – that is, geographically located – so that the presence of more intense localized learning processes (such as clusters of universities and other research institutes) contributes to the emergence and sustainability of RISs. Nevertheless, the circulation of this knowledge is complex and often unstructured (or structured through overlapping constellations of communities of practice), creating space for creative re-combination and innovation in an otherwise spatially fixed RIS. Following Philip Cooke (2001, 950), who first described and coined the term "regional innovation systems," emergent RISs are "non-linear, decentralized and heterarchical" – a description that adheres just as well to the emergence of the Lapita-associated technological system within the Bismarck archipelago as to

the late-twentieth-century biotech industry in Massachusetts, which
he studies. Certainly, it closely resembles the pattern of guild-driven
growth, consolidation, and innovation proposed by Epstein (1998) for
pre-industrial Italian cities, suggesting that this model – though only
developed recently – can be useful in the elucidation of past practices
and landscapes.

However, where archaeologists have investigated the spatiality of
innovation – beyond simple chronological dispersal/diffusion models –
we have tended to lean more heavily on the world-systems framework
first developed in the 1970s (Wallerstein 1974), rather than the more
recently developed RIS model. World-systems approaches explore the
flow of ideas, goods, and people through shifting patterns and long-
term development of global, regional interaction, largely in the modern
period, though archaeologists have applied them to a variety of periods,
peoples, and regions (e.g., Chase-Dunn and Hall 1991; Kardulias 1999;
Kristiansen et al. 1987). The locus of research in this model is interrela-
tion of core, peripheral, and marginal regions. Cores are dynamic centers that dominate the flow of ideas and resources and become hearths
of innovation, while peripheries are net importers of core-originating
innovations and providers of raw material, labor, etc. Margins are only
indirectly affected by activities in the core area, with contact mostly
limited to the use of or exposure to novel and exogenous technologies,
materials, or practices. While the application of this approach to archaeological contexts has provided a powerful synthetic tool for building
elegant grand narratives, such as Andrew Sherratt's (1993) proposed
Bronze Age World-System (Fig. 5.3), it sometimes sits uneasily against
the material evidence of exchange and communication between people
living in pre-capitalist worlds (Hall and Chase-Dunn 1993; Kohl 1987).

As an approach developed to explore the emergence of European
colonial dominance and nascent capitalist relations, world-systems
models often require careful modification to be applied to small-scale
societies with minimal or no established hierarchies and exchange systems in which cosmology and kin relations are as, or more, important
in value creation than the objective worth of raw materials or the desire
for wealth (Peregrine 1996). Indeed, much like recent attempts to apply
globalization models to the ancient past (Vandkilde 2016), it seems to
me that there is little interest among archaeological proponents of this
approach in grappling with the assumptions about asymmetrical power,
inequality, and exploitation between core and periphery – all characteristic products of the very specific social, political, and economic relations
engendered by European colonialism (Lightfoot and Martinez 1995).
Lacking this reflexivity, archaeological world-systems approaches risk
unconsciously retrojecting these political and power relations into the

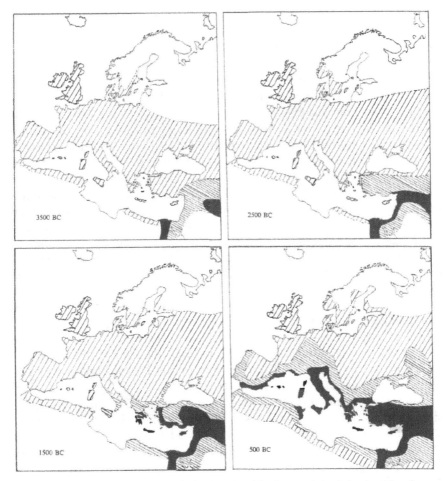

Figure 5.3 A model proposing shifting cores (black areas), peripheries (closely hatched areas), and margins (lightly hatched areas) in Bronze Age Europe and adjacent regions.

past, rather than exploring alternative forms of social, political, or economic interaction.

Although RIS and world-systems models differ in their scale of analysis, they share a distinct conception of the spatiality of innovation, in that it emerges primarily in well-connected core regions, and less so (if at all) in peripheries that either access exogenous innovations in exchange for raw materials (world-systems) (Kardulias and Hall 2008) or innovate only when networked into non-local knowledge networks

(RIS) (Isaksen and Trippl 2017). However, other researchers suggest that this spatial hierarchy is more assumed than real. Geographer Richard Shearmur (2012, 2015) has recently called into question the received wisdom that urban centers are innately innovative. He argues that much of the innovation literature conflates marketing or launch activities with the first introduction of a product or process, eliding the fact that there is no one geographic heartland for innovation. He draws on a wide literature in management and innovation studies to propose that innovation as a process should be unbundled into different phases and industries because some forms of innovation – e.g., in mining or forestry – develop better in rural regions where the industries themselves are located, while others – e.g. wine production – do not require the rapid communication of tacit knowledge that clustering promotes, and might instead benefit from relative secrecy. Indeed, considerable research suggests that permanent spatial proximity is not necessary for many industries to innovate, and intense or focused, episodic gatherings such as trade fairs or conferences are adequate for the transmission of pertinent information (Eder 2019, 121). Shearmur's (2012, S14) suggestion that isolation from information-sharing networks might benefit firms for whom industrial secrets have considerable value is also worth considering in an archaeological context, such as in Crown's (2016) recent work on the transmission of red- versus whitewares in the pre-Hispanic American southwest.

These observations build on, among other things, important work by Canadian geographer Andrey Petrov (2011, 2016), who has studied regional innovation in Canada's remote northern territories. He emphasizes that, in contrast to more city-centric forms of innovation, innovation activities in peripheral areas entail and are strengthened by their "situatedness." By that, he means appreciation and operationalization of local knowledge and relationships as well as engagement with the community. Although he contrasts industrial and post-industrial paradigms of innovation to explain the creative potential in Canada's northern fringe, reading between the lines, peripheral innovation systems were inhibited not by any internal dynamics, but by constraints imposed on these regions by distant governments and industries unable or unwilling to accept alternative modes of innovation and value creation. In other words, a narrow concept of innovation emerging from dynamic core regions is not only an inaccurate reflection of reality but can have the perverse effect of limiting the impact of peripheral innovations.

I highlight this research into contemporary cores and peripheries because I think it mirrors the archaeological discourse. By presenting innovation as apparently limited to geographically core regions, with diffusion or migration as the mechanisms for its outward spread, we under-estimate or efface the creative activities and innovative practices of people living in areas deemed hinterlands, not to mention the complex relationships that connect them to the so-called "core" regions (see Stein 1999). So, in the case study of Lapita, a generation of archaeologists saw a uniform dispersal of a coherent and cohesive social and technological package that degenerated over time, while only recently has attention been paid to the variations in practice, life style, and cultural trajectory between different communities or archipelagos (Cath-Garling 2017; Specht *et al.* 2014). Kienlin (2018) makes a similar case for Bronze Age Europe, arguing that the grand narratives of Mediterranean dependence that characterize our models for this period ignore and render invisible the variability and historicity of local communities in both "core" and "peripheral" areas. Following Kienlin, much of the data in support of these narratives of dynamic Mediterranean core and receptive, emulative northern and western European periphery are divorced from local archaeological contexts and chronologies, an interpretative process that elides cultural difference in favor of normative and linear narratives. He points specifically to the situation of elites in the Carpathian Basin who, within a world-system approach, dwell in the Mediterranean periphery, deriving wealth and status from contacts to the south. Instead, he suggests they played a complex role brokering communication and exchange between neighboring communities, reinterpreting, translating, and passing on amber and other northern European materials while adopting and re-working some Mediterranean valuables. Indeed, rigid models of core and periphery limit our ability to understand how and why innovations spread, since they accept a hierarchical (and very modern) conception of innovation or novelty as innately desirable and valued. As we have seen in previous chapters (and as discussed at some length in Chapter 6), the adoption or rejection of an innovation is rarely a simple, rational, or linear process.

Instead, we might be better served by reconceptualizing peripheries as frontiers: "zones of cross-cutting social networks" (Lightfoot and Martinez 1995, 474), where people with different cultures, ways of life, technologies, and beliefs come into contact and co-exist. In engaging with each other, they necessarily develop transformative boundary objects allowing them to bridge their differences and, in the process, create new social and material forms. Peter Lape (2003) adopts this framework to re-position Island southeast Asia as a dynamic zone of constant interaction, rather than a periphery to more or less distant cores. He argues that the histories of interaction and culture contact in this region resulted in fluid and shifting personal identities, suites of material culture, and networks of interaction, leading to the emergence of innovative social forms, such as the Lapita cultural complex. A frontier approach to these sorts of interstitial regions destabilizes bounded concepts of ethnicity, identity, and power accumulation, allowing us to step away from modern political economies and evaluate supposedly marginal places and peoples on their own terms.

Indeed, heterogeneous networks are highly creative (Mokyr 1992, 328) (discussed further in Chapter 7), and peripheries are often highly heterogeneous – in terms of both social structure and population as well as in their economic and material bases, since these exist in a state of flux between local interests and regional or supra-regional pressures. However, while some frontier inventions might persist over time, creating new categories of object and practice that link disparate communities and allow a shared sense of identity to develop, many others will be ephemeral, limited to the momentary interaction. Whether an invention is adopted and an innovation is dispersed relies on the very cross-cutting social networks described above. Innovation diffusion, in its simplest form, is the transmission of ideas from person to person and community to community. In the small-scale societies of the distant past, particularly those without writing, this process was likely thoroughly entangled with kinship and lineage – social structures that can cluster in space, but also cross boundaries, generations, languages, and social or economic class. In this way, innovations are not tethered to a place from which they diffuse, but are mobile in space and dynamic in form, making a mockery of the rather linear models we archaeologists still like to use to describe their spread.

Beyond migration and diffusion

Archaeologies of innovation tend to bifurcate and pool around either migration or diffusion, two phenomena of longstanding interest to archaeologists concerned with understanding histories of human contact and the dispersal of new or altered things and practices. In general terms, migration models seek to explain the introduction of innovations as a product of mobile population, while diffusion models are vaguer about the mechanism for mobility, often leaning on the concept of *stimulus diffusion*: that is, that people who encountered a desirable new thing or practice adopted it in so far as it fit into their cultural system, needs, or capabilities (Kroeber 1940). In Chapter 3, I introduced the concept of skeuomorphism – cross-material, morphological imitation – in the context of evolutionary models of technological change. Skeuomorphs have also been invoked as proof of stimulus diffusion from core to periphery. Many explanations of apparent skeuomorphs are premised on the idea that people living outside core regions imperfectly reproduced more prestigious, valued, or technologically advanced object types in whatever materials they had to hand. So, the emergence in prehistoric southeast Europe of particularly plastic ceramic forms with swooping curved handles is read by Sherratt and Taylor (1997) as an intentional imitation of molded metal vessels from the neighboring Mediterranean core. Migration, by contrast, puts the onus for change on the innovators themselves, who bring new things with them by either replacing or adding to the material package of the people they encounter on their travels. This is the scaffolding around which many of the models of Austronesian expansion are constructed. Indeed, a migration logic underlies Green's Triple I model of the Lapita cultural complex: it represents the integration of migrant Austronesian-speaking people's way of life with the local practices of the Bismarck archipelago, the innovation of rapid maritime expansion, and the intrusion of Austronesian speakers into areas occupied by Papuan-speaking people.

Migration models fell rather out of fashion in the 1970s and 1980s (Anthony 1990), but in recent years they have come roaring back (Burmeister 2000; Härke 1998). Inspired by early research into prehistoric genetics, population mobility, and cultural transmission (Ammerman and Cavalli-Sforza 1973, 1984; Cavalli-Sforza and Feldman 1981), numerous research teams have applied increasingly precise methods to study the gene-flow of past populations and correlate these with the spread of new technologies, practices, and archaeological cultures (e.g., Haak et al. 2015). These models have come in for considerable critique for their often simplistic usage of archaeological data and past people's identities (Frieman and Hofmann 2019). In a Pacific context, while genetic data seem to confirm major episodes of population turnover, these are not necessarily coincident with cultural changes; and language at least does not seem to change with genetic profile (Lipson et al. 2018; Posth et al. 2018). Diffusion models are equally diverse, including both evolutionary studies of knowledge transmission and the smaller-scale studies of interlocking constellations of communities of practice. Barbara Mills and colleagues have applied social-network analysis to the diffusion of Roosevelt redware in the pre-Hispanic American southwest, charting its uptake and spread through various strongly or weakly connected networks of communities (and, presumably, potters) (Mills et al. 2013, 2016; Mills and Peeples 2019). In this case, they argue that the rapid uptake of this ceramic style in geographically dispersed and heterogeneous communities is likely the result of the high esteem in which expert potters were held – their influence and interconnections promoted a rapid uptake of new potting traditions.

As Mills' work makes clear, the archaeological tendency to distinguish between diffusion on the one hand and migration on the other is flawed. The flow of ideas cannot be separated from the flow of people, as was recognized by, among other prominent migrationists, Ammerman and Cavalli-Sforza (1971), who coined the phrase *demic diffusion* for innovation dispersal linked to progressive population movement. Indeed, as many archaeologists have recently started arguing, we can, furthermore, be considerably more nuanced in how we model people's mobility in time and space (e.g., Beaudry and Parno 2013; Hofmann 2016; Leary 2014). We must move away from linear models of origins and homelands or diffusion "downslope" from cores to peripheries. Social networks cross our analytical categories (i.e., culture group, ceramic type, language family) because archaeological categories are not emergent in the material record but are applied to it to aid in our own analyses. Moreover, they also cross boundaries – both those we invent as heuristics and those we observe in the natural landscape. A limitation of much of the archaeological work on the spread of innovations is that implicit in most models is the concept of bounded peoples, places, and technological systems. Indeed, even frontier and boundary studies that sought to counter the core–periphery hierarchy still accepted that effective and stable boundaries existed between people and communities. If we accept the relational model of innovation I have been developing in previous chapters, then the geography of innovation will be the shifting geography of unfixed and dynamic social relationships out of which they emerge and through which they spread. Boundaries between cores and peripheries, if they even existed in the periods under study, are not inevitable but emanant: constantly created, maintained, altered, breached, and reworked – in other words, mobile and in flux. This is a particularly important point to keep in mind when examining archaeological applications of genetic studies. While aDNA researchers emphasize that there is no such thing as a genetically pure or bounded population, they continue to model their data against the fiction of bounded archaeological cultures and, in so doing, come disturbingly close to reproducing early-twentieth-century nationalist migration studies concerned with finding the homelands of ancestral populations and correlating these with more or less "sophisticated" technological systems (Frieman and Hofmann 2019).

It is also worth considering the temporality of the spread of innovations. While both diffusion and migration describe the broad phenomenon of an innovation dispersing in space – in the former case as the movement of ideas and things, often from core to periphery, in the latter as a product of large-scale human movement – they also have temporality, since they seek to describe and explain change over time. Although widely adopted and highly "successful" innovations tend to follow the S-curve of increasing saturation over time discussed in Chapter 2 (Fig. 2.2), many innovations are not successful, or are only successful in some areas or with some people. There is considerable variety in whether people adopt innovations, even in spatially proximal areas, and this is discussed extensively in Chapter 6. In other words, the temporality of innovation dispersal is also not linear but dynamic. It responds strongly to the articulation of different interpersonal networks, connections between communities, economic processes, and social values – none of which is fixed or immutable. Consequently, we need to be aware not just of the spatial scale of the tools we use to study the dispersal of innovation but also of their temporal assumptions. For example, Mills *et al.* (2016) detail a shifting social network of potting communities from 800,000 sherds of pottery that were produced over about 250 years but are dated in 25- or 50-year increments, more or less a human generation. This temporality gives them insight into a very human-scale process of innovation diffusion. However, aDNA research by necessity explores migration and population mobility through the temporality of gene-flow – a very different scale, as an individual's genetic profile tells a story not about themselves, but about several generations of their ancestors, upwards of two or three centuries of interconnections compressed into one data point. Comparing this to the spread of a potting tradition (e.g., as by Kristiansen *et al.* 2017), which operates on a very different time scale, is considerably more fraught than the archaeologists who have attempted it are yet willing to admit.

Finally, changing patterns of communication, whether written or oral, have the potential to impact deeply the dispersal of an innovation and its reproduction or alteration over time. Much of the research in innovation studies highlights the role that mass media – books, journal articles, radios, newspapers – play in spreading innovations; but clearly these were not available through most of human history. Instead, most innovations have spread through direct contact between individuals, whether they be kin, neighbors, influential members of the community, teachers, or newly encountered foreigners. The impact of specific individuals within a community (or community of practice) on the adoption and dispersal of innovations is unquestioned, but the story of the innovation – its role, function, appropriate usage or deployment, etc. – entails more than a single early adopter to spread. While oral traditions do not reproduce perfectly over time – one of the sources of the "drift" discussed in evolutionary models – they are more than acceptable for sharing and transmitting extraordinarily complex information over very long periods of time. I would point specifically to evidence drawn from Indigenous Australian songs and histories here, which, among other things, have been demonstrated to have recorded accurate astronomical observations and retained memories of major episodes of sea-level change dating to thousands of years in the past (Nunn and Reid 2016). Indeed, the famous "songlines" or Dreaming tracks entangle complex information about cosmology, social relations, land tenure, lineage, important places, and ecological observations, all of which allow routeways to become manifest in the landscape to the person or persons who has the right to and knowledge of their stories and songs (Chatwin 1987; Taçon 2005). Dreaming tracks are an interesting example because they cross-cut languages, communities, and boundaries between different people's countries, and can have considerable geographic extent – even as they vary slightly from one telling or language group to another (Neale 2017). In other words, they too are boundary objects able to bridge distant communities, both metaphorically and geographically. As such, they offer insight into the ways that other complex information, such as metallurgy or agriculture, might have been shared, transmitted, and reproduced generation after generation in community after community.

Elders tell stories to children; as these children grow and age, the stories grow and age with them, increasing the complexity of the information. Upon adolescence or adulthood, when appropriate initiation rites have been completed, they might include secret information as well. In this way, kinship and social relations become fundamental mechanisms to the dispersal of innovations in both time and space. In other words, the geography of innovation is unstable because it is a social geography, and the transmission of information about innovation is a social process mediated by wider cultural norms and achieved through the messy and imprecise articulations of individual people who made up both physical communities and communities of practice.

6

Tradition, continuity, and resistance

Are some societies more conservative than others? Why do some people resist innovation? How do we identify a failure to adopt?

It seems apt to begin our discussion of conservatism by quoting a self-described conservative. William F. Buckley, the renowned twentieth-century American intellectual, journalist, and novelist, described the mission of the *National Review*, a magazine he founded to propound conservative thought, in the following manner: "It stands athwart history, yelling Stop, at a time when no one is inclined to do so, or to have much patience with those who so urge it" (Buckley 1955). Buckley's conservatism is familiar to any follower of United States politics: it demonized all forms of collective action, particularly communism; resisted centralized government power at home and, in the form of the United Nations, abroad; reviled the Civil Rights and Feminist movements; and was a muscular proponent of libertarian principles and capitalism.

Although, in conformation, this articulation of conservative principles clearly derives directly from the social and political context of the post-World War II United States, in its stated fears of "energetic social innovators, plugging their grand designs" it is a direct descendant of a much older intellectual approach to innovation. As Godin (2015a) has thoroughly drawn out, ancient-through-early-modern Christian intellectuals explicitly framed innovation as subversive, heretical, and posing an innate and existential danger to the social order. Yet, this definition of innovation centers almost in its entirety around *social* innovations, leaving technological innovations largely unremarked upon. Whilst social conservatism is a thriving political movement, technological conservatism is widely reviled, being explicitly problematized in scholarly literature (Moore 1991, 1995) and viciously caricatured in popular media. The face of technological conservatism is old, out of touch, out of date and waving a slipper at those damn kids on his lawn – like Grampa Simpson, he probably also wears an onion on his belt.[1]

This disconnect in our discourse between social innovation/conservatism and material or technological innovation/conservatism is one that has served to obfuscate the ways in which conservatism operates in society. Moreover, it is grounded in a false distinction, since the material world is itself bound up in the social, being wholly implicated in human social practices and interpersonal relationships. Thus, the rejection of new technologies or technological practices is as much a social decision as are the rejection of shifting definitions of marriage, or challenges to traditional gender roles and identities.

So far, I have attempted to explore the creation and diffusion of new things and ideas. I have examined why innovations are adopted, how they develop, and the ways in which they are communicated, but I – like many others – have only gently touched on the topic of innovations that are not adopted, are rejected, or fail. There is, in fact, little in the way of a standard approach to the opposite of innovation, perhaps because the rejection of innovations is rarely embraced in the present day. Yet most innovations *are* rejected by some, if not the majority, of the people who come into contact with them. Arguments have been made, and more recently rejected, that pre-industrial societies were innately conservative: that is, that they largely rejected and resisted innovations unless these were forced on them (by environmental or demographic pressure, by outsiders invading, etc.). However, innovation failure is common in the present – long-running jokes about the Edsel and Betamax home videos persist even though these innovations failed before I was born – and we claim to live in a deeply innovative and innovation-welcoming culture. In this chapter, I tackle the thorny question of conservatism and innovation resistance, from how we can identify these behaviors to a better understanding of conservatism as a dynamic social practice closely bound up in the negotiation and relations of power among individuals and groups in a given community or society. The key to much of this discussion is "tradition," a slippery concept that encompasses both specific ways of doing or living and a structure of intergenerational knowledge and practice. Identifying non-change or a reluctance to innovate in archaeological contexts is not straightforward, but I find later prehistoric Cornwall – the area of my own current fieldwork – to be a really apt example.

Maintaining traditions and emphasizing local identities in Iron Age and Romano-British Cornwall

The end of the first millennium BCE and the start of the first millennium CE saw considerable upheaval and change across much of the European continent. In Britain, the impact of contact with the Roman

Empire was felt well before the first attempted invasion as new practices, tools, foods, and ideas were adopted widely, particularly in the southeast. With the Roman invasion in the first centuries CE, we see a major shift in settlement patterns, with the emergence of proto-urban and urban centers; a re-organization of the rural landscape to suit the Roman villa system of land management; and the introduction of major infrastructure, such as roads and extensive boundary defenses. The way of life that emerged in the early first millennium CE was different from that led by earlier generations and reflected a widespread adoption of Roman material culture, cosmology, and practices that were blended with pre-existing ideas and ways of life into a uniquely Romano-British hybrid (Eckardt 2014; Mattingly 2006). However, this pattern of rapid adoption of new practices and identities and of marked change was not universal across the British Isles. In Cornwall, Britain's westernmost peninsula, unambiguously Roman sites are few, comprising just three briefly occupied small forts, a rather atypical villa, and five milestones located in close proximity to ore sources. Only about 1,000 individual finds of Roman materials have been recorded in Cornwall – of which about two-thirds are stray finds of coins – and a handful of apparently Roman-style burials have been excavated (Nowakowski 2011). It seems fair to suggest that most people living in Cornwall in the early first millennium CE were not living a broadly "Roman" way of life (Henderson 2007, 220–9). Instead, to an archaeological eye, Cornish ways of life appear to have remained much as they were in the centuries before the Roman invasion of Britain (Cripps 2006).

Cornish people in the first millennia BCE and CE led largely rural lives. Small, enclosed farmsteads of less than 1 ha, locally referred to as "Rounds," are the primary archaeologically visible settlement type from about 200 BCE through to c. 600 CE. Unenclosed settlements seem to have dominated before this point. Thousands of likely Rounds have been recorded by the Ordnance Survey, by archaeologists, and through aerial photography; but few have been excavated (Lewis and Frieman 2018). A conservative estimate suggests that there is a Round every 2–4 km² in Cornwall – and in some areas they are much more closely spaced (e.g., Rose and Preston-Jones 1995). Whilst it is unlikely that these were all concurrently occupied – or even that they were all settlements – those few that have been excavated have yielded evidence for extended periods of occupation and multiple phases of reconstruction, suggesting that they were not frequently abandoned. A variety of Round morphologies is known, from simple, univallate circular enclosures to bivallate forms; rectilinear enclosures; and enclosures with large, often irregularly shaped annexes that may have been used for stock keeping, but also sometimes contained house foundations (Fig. 6.1) (Lewis and Frieman 2018).

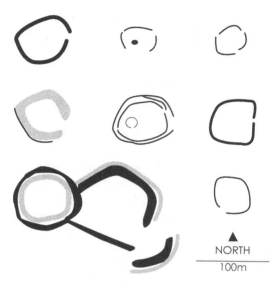

Figure 6.1 A selection of plans of Cornish enclosures likely dated to the Romano-British period. Ditches are black, banks grey.

The houses themselves are few in number. Between two and six houses seems to have been normal, though one or more of these would likely be used as outbuildings, stables, or craftworking areas at any given time. They were typically more or less circular – a very old style of architecture in the southwest of Britain, dating back to at least the first half of the second millennium BCE (the local Early Bronze Age); but some were distinctly oval. Oval buildings have been interpreted as a style of vernacular architecture unique to Cornwall during the Romano-British period that drew on much older building techniques, craft knowledge, and ways of life (Quinnell 2004, 205).

There has been little discussion in Cornwall about the nature of these communities, except to suggest that they represented the wealthier, landowning classes. Based on the small size of most of the settlement Rounds and the low number of structures within, it seems likely that these were occupied by extended family groups, and their long periods of occupation probably reflect carefully protected, familial inheritance rights. Of interest is that, to date, there is no evidence for Round construction after the second century CE, even though they were carefully maintained, slightly altered, and remained occupied, in some cases, for centuries after that. Clearly, they were a valued form of settlement and one that retained its social and political significance for generations. Some portion of the population also probably lived in unenclosed settlements that are only just being identified archaeologically (Young 2012). These groups are typically considered to be the poorer agricultural workers, beholden to the Round-dwelling landowners, but, consistent with small-scale agricultural communities known historically and anthropologically, they were also likely part of the same complex kinship networks as those who dwelled inside enclosed settlements. The web of obligation between them would have pulled in both directions.

The best known and most thoroughly studied Round is the site of Trethurgy, which was nearly 100 percent excavated in the 1970s but only published in 2004 (Quinnell 2004). Based on radiocarbon dates, careful observations of archaeological stratigraphy, and analyses of the ceramic and other artefactual assemblage, Henrietta Quinnell has suggested that the enclosure site was built and occupied over at least 500 years. An initial enclosure, perhaps used for stock keeping, was built in the first century CE; and, between 150 and 550 CE, it was a settlement site, with around nine phases of occupation linked to re-building and alterations of the exterior embankment and the internal structures. These structures consisted of a number of round and oval buildings, often re-built over subsequent phases. Each of the larger houses could have housed a family group of up to twelve or fourteen people, so the whole population of Trethurgy was likely no more than a few dozen. The agricultural practices identified during the centuries of its occupation represent a continuation of Iron Age subsistence practices. A small kitchen garden may have been present within the embankment, but there are indications of a surrounding field system in which crops, including spelt, emmer, and barley, could have been grown and animals pastured (Jones 1991). Nearby upland areas were probably exploited for summer grazing. Most craft activities seem to have taken place communally in the open spaces between the buildings, with areas for grain storage and cereal preparation having been identified during excavation. Quinnell (2004, 232–3) believes the inhabitants to have been largely self-sufficient, practicing a variety of crafts from leatherworking to stone carving to smithing as necessary.

Nevertheless, the Trethurgy community was not isolated. Not only were there other Rounds in close proximity, but markets and rituals would have likely brought communities together to exchange goods, negotiate marriages and other alliances, pay taxes to the regional administrative center, and jockey for status. Certainly, the Trethurgy inhabitants had access to tin and smelted iron, which they probably acquired through exchange. They drank a small amount of wine – as evidenced by a handful of amphora sherds present on the site – and had some Roman tableware, including a couple of glass vessels, samian ware, and Oxfordshire ware. These special vessels show distinct signs of curation, and the samian ware in particular seems to have been deposited several centuries after its primary period of circulation. Two brooches dated typologically to the second century but found in a fourth-century context support the idea that these may have been carefully curated heirlooms. Quinnell suggests that the brooches may have served as badges of status. Together with the heirloom eating vessels and careful maintenance of a centuries-old enclosure, the image of a community very interested in maintaining traditional ways of life and looking towards the past, perhaps as a source of status, authority, or rights to land, begins to come into focus (see Cripps 2006).

Indeed, the ceramics from Trethurgy and other contemporary sites offer a further line of evidence for a continued investment in traditional Cornish practices. These were predominantly formed from gabbroic clays over an 18 km^2 area of gabbro bedrock outcrops on the Lizard, a peninsula in the far west of Cornwall. These clays are easily accessed, and experimental archaeology suggests that the vessels made from them are highly resistant to heat shock and open firing, making them ideal cooking vessels (Quinnell 1987, 2004, 109). This clay source was used throughout prehistory, and sherds of pottery with characteristic white feldspar and grey augite inclusions have been recovered on Cornish sites dating from the fourth millennium BCE through to the first millennium CE. Of the 550 minimum vessels identified through analysis of the Trethurgy ceramic assemblage, 450 were made of gabbroic clay. In other words, during 400 years of occupation at Trethurgy, over 8 in 10 vessels used were made from clays sourced in one small corner of Cornwall. This pattern is consistent with assemblages from other Late Iron Age and Romano-British Cornish sites (e.g., Appleton-Fox 1992, 94–113; Carlyon 1987). While the organization of potting must have changed over the 5,000 years in which Cornish people specially sourced gabbroic clays for pottery, the fact that this single raw material was so highly sought after for so long speaks to the longstanding preference for tradition in Cornish prehistory.

Ceramics made from gabbroic fabrics are visually and texturally distinct, with numerous inclusions of white feldspar that protrude though the surface of the clay. Their look and feel would have formed part of the sensoria of cooking and consuming food over countless generations, and their value should probably be understood in that context. In other words, the longstanding use of gabbroic pottery is not the product of technological stagnation or unthinking conservatism.

To some extent, the distribution of the Rounds in the Cornish landscape may speak to a sense of shared identity in the face of outside contacts. It is well recognized that many of the Rounds are often located on sloping land or spurs, rather than on the crests of hilltops like the earlier hillfort enclosures. Rounds were not always placed in topographically impressive locales, a situation that contrasts somewhat with their identification as higher-status sites. So, for example, we might look at the site of Hall Rings near the modern village of Pelynt (Lewis and Frieman 2017, 2018) (Fig. 6.2).

Hall Rings is a complex Round comprising a primary circular enclosure bounded by multiple banks and ditches and a D-shaped annex adjoining its entrance on the southeastern side. The site is located on a southeast-sloping spur flanked by streams flowing into the West Looe River, which would have been navigable in prehistory. An outer embankment several hundred meters upslope from the main enclosure is thought to be contemporary with at least some phase of its occupation. Topographic and geophysical surveys show that outer banks and ditches of the main enclosure were more numerous and more massive on the upslope, western, and southwestern sides of the enclosure and that some sort of elaborate entrance architecture had likely been constructed at the entrance to the inner enclosure from the annex. Several structures were identified within the annex and inner enclosure, although the contemporaneity of these cannot be judged.

Based on the complexity of the enclosing banks and ditches and the number of interior structures, it seems likely that Hall Rings, like Trethurgy, was occupied for a considerable period of time. Unlike Trethurgy, however, although field divisions were present between the enclosure and the outer embankment, these likely relate to late or post-medieval land divisions rather than to the prehistoric occupation of the site. Instead, we believe that the outer embankment served both to control and to channel access, including visual access, to the inner enclosure, as well as containing or protecting local pasturage. Anyone approaching the settlement at Hall Rings would first have to navigate around the outer embankment and then commence a long and exposed downhill walk towards the multiple banks and ditches on the inner enclosure's western side. In order to enter the enclosure, one would have to walk around these monumental outerworks, and then move through a narrow entrance into the southeastern annex and uphill into the elaborated entrance to the main enclosure. In short, at Hall Rings, we are looking at an extensively elaborated small-scale settlement that is, nonetheless, shielded from view and hard to access.

In her study of Trethurgy, Quinnell (2004, 213) identified the significance of "an imposing gateway" to the inhabitants of Round settlements. Rounds would have been entered not only by their inhabitants but also by visiting outsiders, kin, and neighbors. They may also have been sites of small-scale exchange. Almost inevitably, these gatherings would have served as venues for status negotiation through displays of wealth

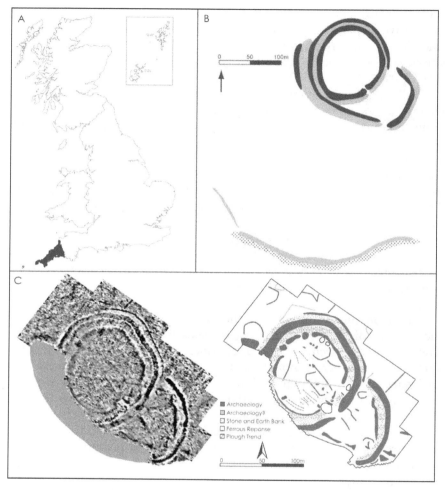

Figure 6.2 Hall Rings enclosure; (a) map of Britain with Hall Rings indicated as white dot; (b) topographic survey of Hall Rings enclosure and outer embankment. The stippled area is Holloway to West Looe River; (c) results of magnetometry survey of Hall Rings (l) with interpretation (r).

and power. Elaborated entrance gates, as well as complex and guarded approaches as seen at Hall Rings, suggest that there was an interest in controlling who could enter a Round and how, and considerable investment in differentiating between residents and outsiders. Interestingly, Hall Rings was not at all an isolated site. Numerous other enclosures have been recorded in the near vicinity, most strikingly the elaborate round at Bake Rings, just 2.5 km away on the other side of the modern village of Pelynt. Bake Rings also consists of a small circular enclosure with a large banked-and-ditched annex located on the downslope of a hill. Here, too, we find an elaborate entry gate – although in this case it is the entry to the annex, rather than to the small, circular enclosure, that has been elaborated – and ostentatious outerworks, consisting of an enormous ditch many meters across and over 2 m deep, enclosing the annex. If these sites were occupied contemporaneously, these excessive exterior constructions, complex approaches, and imposing gateways may reflect local status competition – a case of "keeping up with the Joneses."

However, despite this elaboration, neither site was located in a visually imposing locale, nor were they visible from nearby navigable waterways or hypothesized local overland routeways (Frieman and Lewis 2016). This suggests that only locals would have known where to find these settlements and only locals would have known how to approach them. Just as the extensive embankments indicate an interest in distinguishing those who belong and those who do not, the whole occupied landscape appears to repeat this pattern, having been laid out with local interests in mind and representing political, economic, and social competition between and among insiders. This sort of landscape organization likely reflects the underlying political and economic realities of the time: the landscape was divided among powerful extended kin groups who made their residences in Rounds, and who would have competed with each other for status and prestige on a local rather than regional level, forming a relatively stable heterarchy. The small numbers of Roman and other exotic materials suggests that access to outside goods and materials were of secondary interest to more local concerns, chief among these probably being the negotiation of alliances that allowed for the seasonal movement of stock between the densely occupied lowlands and communal, upland grazing areas to the north (see Herring 2009). Here, the lack of a centralized authority or clear hierarchical political structure probably contributed to the maintenance of the much older ways of life discussed above. Any major divergence from the carefully balanced network of kinship, status, and land tenure represented by the dense distribution of the Round settlements would have made participation in these negotiations difficult-to-untenable, since it would have made one an outsider and this was, undeniably, an insider's landscape. In Romano-British Cornwall, enacting traditional practices – including the invention of a new form of settlement, the Round, based on pre-Roman kin and land-tenure networks – served as a sort of infrastructure of social solidarity,[2] allowing a distinctly Cornish identity and way of life to persist even in the face of the Roman army.

Spotting innovation rejection and technological conservatism in the past

In the contemporary societies studied by sociologists and anthropologists, it is generally easy (on the surface – hold that thought, we will return to it below) to identify innovation rejection and technological conservatism. Classic sociological studies of rural communities have focused on, among other innovations, the patterns and speed of the varying adoption processes for chemical fertilizers, new crop varieties, and new farming technologies (all discussed at length by Rogers [2003]). For example, Ryan and Gross (1950), in charting the relatively rapid adoption of hybrid corn varieties in early-twentieth-century Iowa, were able to observe the varying levels of resistance this innovation faced as it spread through local communities, and the specific personality traits and decision-making processes of individual farmers. Their results echo the stereotypes about technological conservatism noted above, namely that later adopters – those who resisted the adoption of hybrid corn for a number of years and only adopted it after a slow process of familiarization and small-scale trials – were, in general, older, less educated, less wealthy, and less cosmopolitan. Ryan and Gross note particularly that farmers more embedded in rural communities who had spent less time engaging with urban institutions were among the most resistant to the new hybrid corns, making clear that the rural social networks themselves could be perceived as innately conservative in this specific case. However, archaeological insight into these sorts of individualized decisions and small-scale, short-lived adoption processes is typically limited because of the nature of the materials we study (and increasingly so the further back in time we wish to investigate).

One approach to this problem might be to explore industries or domains, such as food production and cooking, that tend towards conservatism. Here, continuity of practice may relate to the fact that cooking and eating are highly emotive activities, and ones that are often central to the formation and maintenance of distinct ethnic identities (Arvela 2013; Brown and Mussell 1984). So, returning to Cornwall, we might see the longstanding use of gabbroic pottery as part of an established and maintained preference for a traditional, sensuous experience of food preparation and consumption. Indeed, ceramic forms changed significantly during the Iron Age and Romano-British period, with new individual bowls and dishes being introduced in the first century and large serving platters in the fourth and fifth centuries CE, suggesting shifts in the practice of food consumption. Nevertheless, these vessels were also largely made from gabbroic fabrics. In fact, even though pottery production in most of the British Isles became much rarer in the early medieval period, gabbroic pottery forms continued to be produced and used in Cornwall into the eleventh century CE (Thorpe 2011).

Certainly, the longevity of the gabbroic pottery, even in the face of new vessel forms, potting technologies, and styles of consumption and food preparation, attests to a very long history of highly localized eating and cooking traditions that would have served both to bind these small communities together through shared practices and to allow them to distinguish themselves from outsiders. Returning to our discussion of teaching and learning from the previous chapter, it is possible that this technological continuity reflects how these technologies were taught, presumably by parents or older kin to younger family members. In particular, the appropriate choice of raw materials would have implicated not just explicit teaching activities, but also wider (kin?) connections, as gabbroic clay was not local to all households. There is perhaps also an interestingly gendered element at play here, as both ceramic production and food preparation are frequently women's labor in small-scale societies. Perhaps this experience of food consumption reflected marriage patterns in which daughters remain in the familial residence whilst sons move in with their wives. Alternatively, we may be seeing some element of power negotiation, as food consumption and the transmission of cultural knowledge around food and its preparation have been regularly identified as domains in which women, particularly older women, are able to gain power and authority in strongly patriarchal societies (Bajic-Hajdukovic 2013, 57 with references).

This narrative of conservatism or continuity is complicated for archaeologists by the classic "absence of evidence" paradox. That is, are we not finding traces of practice/technology/material X in a given area at a given time because the people living there/then did not have/want it, or, are we not finding those traces because they were not deposited in ways that are archaeologically visible, they existed in such minimal quantities that they are likely unrecoverable without a real stroke of luck, they have decayed or otherwise been destroyed over time due to taphonomic and anthropogenic processes, etc? This is one reason that firsts and earliests have such a pervasive and alluring currency among archaeologists and the public: by dint purely of existing and of having been identified and dated, they myth-bust. They create new and older histories – of technological practice, of religious belief, of territorial occupation – that automatically debunk accepted and widely promulgated narratives that were, themselves, based on the idea that the absence of evidence was meaningful, in that it indicated an absence of people, practices, materials, etc.

There are myriad examples of this paradox, and most have proved fruitful ground for decades of archaeological debate. Okumura and

Araujo (2014), for example, observed a striking and very extended stability in the form of stone points made by hunter-gatherers in southern Brazil between about 12,000 and 7,000 years ago. This conflicted with their evolutionary (selectionist) framework for understanding the archaeological record, as change is assumed to be constant by selectionist models. Instead, they propose a series of possible explanations for this apparent conservatism. They hypothesize that population size may have been so small that few innovations were introduced or, if they were introduced, there would not have been many people to adopt them, and in addition that there was a strong cultural value placed on the faithful reproduction of point forms, leading to conformist transmission. Where they do observe variation, they attribute this primarily to increasing population size over time.

By contrast, to return to the story of the adoption of metal in Europe (discussed in Chapter 3), although meaningful and observable change occurred, this was slow and punctuated. For example, early-fourth-millennium BCE sites in southern Scandinavia (the local Early to Middle Neolithic) have yielded only a small number of copper objects (Klassen 2000) and scant evidence for local metal production (although see Gebauer et al. 2020). Current archaeological consensus is that limited metal objects were in circulation in Scandinavia during the Neolithic, but few in the area knew or were interested in developing the technology to produce them locally. After around 3200 BCE, copper objects disappear from archaeological contexts entirely until the later third millennium BCE, when metal, and metal technology, re-emerges in Scandinavia as part of the Bell Beaker package (Vandkilde 1996). This apparent resistance to metal technology is in contrast to other domains in which southern Scandinavian people of this period proved themselves to be very open to new ideas, practices, and technologies, such as agriculture, the construction of ritual monuments, and shifts in settlement practice linked to long-house construction.

It has been suggested that copper deposition ceased in these areas because of a disruption in the supply or availability of the raw materials (Ottaway 1989), but this suggestion would seem to imply that people wanted copper and could not access it. Although, in the present, we view metal as a better and more valuable technology than its predecessors, we cannot make this assumption for past people, who may only rarely have come in contact with singular metal objects. Alternatively, these metal objects may have remained in circulation but not been deposited in archaeologically visible ways (Klassen 2000, 273). Again, this is possible, but seems unlikely as scientific analysis shows that the metals circulating in the fourth and third millennia BCE were distinct from each other, and earlier metal compositions have not been documented in later contexts

(Klassen *et al.* 2007; Krause 2003). Finally, we are left to ask whether the infrequent deposition of copper objects in southern Scandinavia and the lack of interest in developing a local copper industry in the fourth millennium BCE are better understood in light of the social value, role, and significance of copper. Still, disentangling whether the new material was actively resisted or whether its disappearance at the end of the fourth millennium reflected a more passive disinclination to seek it out is unclear. I would perhaps suggest that one answer to this question might be found in looking at the networks of contact and communication that criss-crossed Europe at this time. If, as has been discussed, kinship networks were a key vector through which metal objects and metallurgical knowledge were disseminated, then perhaps copper was only valuable in so far as the kin ties through which it was acquired were. In other words, disambiguating between the absence in the archaeological record of a given innovation and its conscious rejection by people is not an obvious task, nor is it one with clear-cut answers.

Moreover, there is a longstanding tendency to assume conservatism to have been the status quo, leaving the occasional innovation as the only phenomenon to observe (e.g., Walsh *et al.* 2019b, 66). Past people are imagined as being like the disconnected rural farmers discussed above: suspicious of new ideas and resistant to change. The whole culture–history framework, developed to give chronological, spatial, and social structure to the complex and fragmented archaeological record, is based on the belief that ancient cultures were internally homogeneous over long stretches of time and resistant to change from inside (Watson 1995, 684). Stimulus from without was conceived of as the primary (or even only) means by which change (social or technological) could occur in the past (Vander Linden 2011, 26–7). Whilst most archaeologists would no longer consider themselves culture historians, the ideas of an innately conservative past are still very much present in contemporary archaeological discourse (e.g., Snell 1997, 119–21). Indeed, we still use the typo-chronologies and cultural sequences developed by culture historians, and we still reconstruct parsimonious models of invention that envision fewer rather than more individual moments of invention and phases of successful diffusion and adoption (as discussed, for example, in the case study of early agriculture presented in Chapter 2).

Of particular concern is that these parsimonious models implicitly reproduce colonialist frameworks in which specific societies (often Eurasian) with the trappings of "civilization" (literacy, civic architecture, complex political economies, etc.) are innovative and technologically advanced, while peripheral societies are the less complex (and, thus, less civilized), conservative, passive adopters. Certainly, following Montón-Subías and Hernando (2018), the historical and archaeological discourse privileges *change* – both social and technological – over continuity because of its origins in a particularly Eurocentric and masculine strand of Enlightenment thinking in which change and progress are the hallmarks of civilization. Moreover, having developed within the context of European colonial expansion, it explicitly positioned dynamic European societies as more advanced, more valuable, and more moral than unchanging, stable, conservative, or "backward" non-European societies, as we saw in Chapter 1. Although, in recent decades, archaeologists (like so many others) have grappled with the legacy of racist and colonialist beliefs woven into the foundations of our discipline, we are still developing methodologies to study the past without falling back on common wisdom derived from these pernicious modes of thought. This historical background, added to the inordinate valorization of innovation in contemporary western society, forms an enormous roadblock to our ability not only to recognize conscious acts of rejection or conservatism in the past, but also, once recognized, to puzzle out their internal logic.

Unbundling conservatism

Until now, I have elided much of the difference between phases of resistance to innovations; specific examples of failures to innovate or adopt innovations; and more widespread disinclination to innovate among a social group, political class, etc. At this point, I believe it is worth unbundling these, as they do not necessarily all equate. An individual act of resistance, or resistance to a specific genre of technology or practice, does not mean that a person or group of people is inherently anti-innovation. Indeed, claims that a specific group (e.g., unions, uneducated people, poor people, Indigenous people) is predisposed to reject innovations rarely hold up in the face of more nuanced data, since episodes of resistance are usually linked to specific innovations, situations, and social constellations. For example, novel things may not be able to be integrated into the established social and human-technological networks, or they are perceived to be grossly harmful to these. Moreover, example after example shows that these sorts of narrative generalizations are often deployed by dominant social and political groups to impose new technologies or practices on those who are more marginal by dint of wealth, race, employment, etc. This is particularly true when thinking about the ways non-Europeans reacted to and resisted elements of European culture and technology. Therefore, in approaching the concept of a more generalized social conservatism, particularly when looking at the past, we must be cautious not to reproduce the power dynamics of the modern period.

Nevertheless, as every study of innovation, and innovation diffusion, has made clear, some element of the population often resists innovations, and patterns of resistance build up over time, creating the appearance of groups or regions where conservatism, rather than innovativeness, is the more dominant social mode. In previous chapters, we have discussed Hägerstrand's (1966, 1967) geographical approach to the diffusion of innovations. Hägerstrand's work, like Rogers', emphasizes the importance of influential community members in the diffusion of innovations. His maps of this process show expanding circles of innovation adoption around central nodes of early adopters (Fig. 6.3). In this, they are very much like archaeological distribution maps that show the distribution of artefacts or site types and their increasing (or decreasing) quantity over time. Here, though, what interest me are the vast empty areas around these nodes; those are the areas that did not adopt – that is, that either passively allowed the innovation to fail when introduced, or actively resisted it. In archaeological contexts, these are the areas that we usually avoid discussing, or that we discuss in terms heavily qualified by "as yet" and "so far" and "to date" because our dataset is so fragmented that an extrapolation from a lack of finds is highly dubious. However, I would argue that acts of resistance and innovation rejection are not an unknowable "black box" or an empty place on our maps, but derive from the same social and socio-technical interactions as the adoption and dispersal of new ideas, practices, and technologies.

A first step in understanding conservatism is recognizing that it, like innovation, is neither simple nor unitary. Different peoples may be conservative in different ways. In some places and times, we might see the regular rejection or resistance to specific social practices, whilst new technologies are welcomed avidly; and in others, both new social and technological practices might be outright rejected. Others go further, arguing that even resistance enfolds multiple processes within itself. Martin Bauer (1995) contrasts active resistance with the more passive avoidance of innovations; and Hernán Thomas et al. (2017) expand on this, arguing that active and passive (or "non-working" and "working," in their slightly confusing terminology) resistance are further divided into resistance activities meant to communicate resistance within one's own society and against outsiders. For example, a form of active (non-working) resistance carried out with one's own community as intended audience might be the generation of rival goods and systems, while that carried out against outsiders could include the destruction of goods and systems.

Moreover, if we accept social constructivist models of technology (Bijker 1995; Pinch and Bijker 1987), then non-innovation or innovation failure is actually a regular feature of the development of any new technology, but the successful adoption of a given innovation conceals the many points at which elements failed to be adopted or were outright

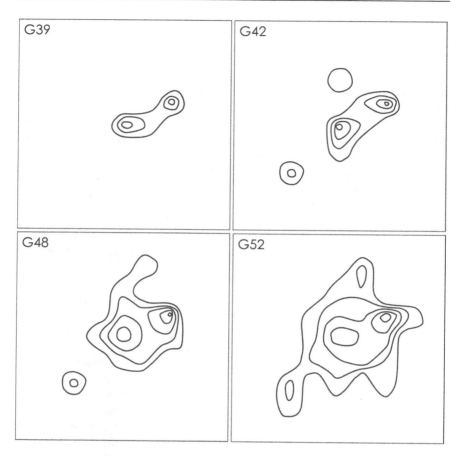

Figure 6.3 Several generations of Hägerstrand's "Model IIIb" of the hypothesized spatial diffusion of agricultural innovations.

rejected. Pinch and Bijker's example of the invention of the bicycle (discussed in Chapter 2) illustrates the regularity of innovation failure and rejection within the messy and non-linear process of invention and widespread adoption. They make clear that the failure of some early bicycle forms was not caused by inferior technology or functionality but resulted from the competing desires and interests of specific groups of people wanting to use them until the stages of stabilization and closure effectively re-wrote the history of the bicycle, making the failure of all other forms seem inevitable rather than the product of long-term social negotiations and complex, competing value systems. This example serves to remind us, first, that the failure of an innovation cannot be judged purely on its technological or functional merits, as these are intrinsically enmeshed in and, thus, influenced by and interpreted through the lens of their wider social, political, economic, and technological spheres. Second, it makes clear that people reject innovations for a wide variety of reasons, and that even within a single society – in this case nineteenth-century British cyclists – different individuals or communities may very well make quite different choices about whether to adopt or reject a given innovation.

When innovation isn't the right choice

Innovations are not universally accepted; are often resisted; and, as discussed in previous chapters, their adoption is often conditional on a complex and non-linear process of experimentation, adaptation, and engagement with friends, family, neighbors, and community leaders. Until recently, however, little effort has been made to develop complex narratives of non-innovation (though see the excellent essays edited by Godin and Vinck 2017). Functionalist and environmentally determinist models still hold far too much sway in archaeological understandings of how and why people began (or ceased) to do new things. This is visible, for example, in many of the assumptions underlying current debates about the peopling of Sahul – the large landmass that includes continental Australia, Tasmania, New Guinea, Seram, and adjacent Indonesian islands. We have well-dated archaeological sites from Indonesia to Tasmania placing human occupation of Sahul between 40,000 and 50,000 years ago. Recent archaeological work at Majedbebe (Northern Territory) suggests that the earliest colonization of Australia may have taken place over 65,000 years ago, with a small population tightly clustered in the tropical north for millennia before they rapidly expanded across Australia (Clarkson *et al.* 2017).

Almost no attention has been focused on why people might have chosen not to leave their home region, with the assumption being that obviously the population would expand outwards as soon as the climatic and environmental conditions made this favorable. Human mobility in these models is described in terms of environmental risk, climatic amelioration, and population pressure (Veth 2005; Williams *et al.* 2015). Although complex social networks and symbolic activity are recognized even among the earliest settlers (Balme *et al.* 2009), this is secondary or even subsidiary in discussions of human expansion across Sahul. For example, Williams *et al.* (2015) employ only economic optimization models to explain why people would or would not choose to move beyond their home regions into more arid parts of Australia. Yet, this sort of simplistic model in which environmental push–pull factors were the only criteria influencing the choice to remain in one's traditional homeland or to leave it can be instantly disproved by the rich ethnographic records and oral histories testifying to the complex and deeply social connection contemporary Aboriginal people have with the Australian landscape, and particularly with their own Country. Moreover, genetic studies hint that Indigenous connections to Country have deep roots (Tobler *et al.* 2017), implying that, over 50,000 years and several phases of population expansion and retraction, a highly conservative sense of origin and relationship to place has been retained. This militates against models of humans more or less rationally and passively, allowing the climate to shape and direct their occupation of the landscape as well as the assumption of constant expansion built into most of the archaeological studies of the peopling of Sahul. In fact, interrogating apparent conservative practices more closely (an all too rare occurrence) tends to illuminate that underlying these are complicated social factors entwining kinship, cosmology, and the balance of power between competing individuals or groups, among other factors.

To contrast with the overly environmentally deterministic models of Pleistocene hunter-gatherer mobility, we might look to Crown and Wills' (1995) work on the initial, apparent failure to develop ceramic containers in the American southwest despite centuries of familiarity with ceramic technology. Although ceramic figurines were produced in this region from *c.* 800 BCE, it was not until over a millennium later, around 400 CE, that ceramic technology was used for the production of pottery vessels. This delay in development controverts the normative archaeological assumption that pottery vessels were functionally superior to earlier organic vessels and would be immediately preferred should the technology be available to produce them. Moreover, pottery production did not develop alongside agricultural domestication, which emerged in the second and first millennia BCE, but was invented or adopted centuries later. Instead, Crown and Wills look to ethnographic research into the gendered division of labor to suggest that pottery production lagged behind subsistence agriculture in this region because it was time-and energy-consuming and interfered with farming and gathering activities. A complex web of causes, including increased sedentism linked to a greater dependence on cultigens, shifts to social structure potentially favoring earlier weaning, and the domestication of new crop species all contributed to the emergence of ceramic vessel production *c.* 400 CE. In other words, a constellation of interrelated social, economic, and environmental factors all affected the ways that people – probably mostly women – decided to use their time and energy to engage with ceramic technology, and the development of new vessel forms was not a viable or desirable option.

Disruption, power balances, and resisting innovations

This complexity in the decision-making process behind innovation failure or rejection is evident well beyond small-scale societies or archaeological case studies. For example, Kline and Rosenberg's (1986) germinal overview of complex innovation processes draws on a variety of contemporary product and process innovations – from jet engines to containerization – to argue that economic and technical models of innovation that emphasize functionality and linear progress are flawed. Arguing for a model of high-tech innovation in which technical and economic processes are intertwined both with each other and with social factors, they highlight the value of old technologies and the difficulty of introducing innovations. For example, they point to the resistance of unions to the introduction of standardized shipping containers for overseas transport. Although the technical modifications to ships and dockside equipment associated with containerization were minimal, and the eventual improvements to productivity and cost savings substantial, the innovation itself was perceived (by the workers who would be displaced by this more mechanized process) as disrupting a stable and beneficial social and economic system. Thus, it was resisted.

Disruption is, of course, a central tenet of the contemporary world of innovation promotion. Coined by Clayton Christensen in his widely read book *The Innovator's Dilemma* (1997), disruption is posited to be the force by which new technologies and businesses succeed and established ones fail. The concept has since metastasized, being applied to industrial developments, high technology, and perhaps most controversially the education sector. Although the desirability of disruptive innovation is widely accepted in popular writing about technology and in academic business programs, critiques have focused on the post hoc identification of disruption within a given market; the concomitant diffuseness of the terminology itself; its conflation of different categories of innovation; and its poor record as a source for predictive modeling, despite Christensen's claims (Danneels 2004; Lepore 2014; Markides 2006). Fundamentally, as Jill Lepore (2014) identified in her critique of the concept, this is a reformulation of Schumpeter's (1943, 81–6) long established, and much more cynical, idea of *creative destruction*. Schumpeter, inspired by Karl Marx, argued that the competition at the heart of capitalist economies requires a constant flux in which new ideas, businesses, technologies, etc. destroy their predecessors as they emerge, are taken up, and made profitable. Through the process of creative destruction, innovations emerge, but established practices and relations of production are reduced to ash, leading, in Schumpeter's view, to the inevitable crumbling of the whole capitalist system and the political and social institutions reliant on it (139–42).

It is this social element on which I wish to focus in order to draw this digression back to the topic of conservatism and the rejection of innovations, since, when asking why an innovation or a series of innovations might be rejected or resisted, it seems pertinent that we think about what in particular is being destroyed or disrupted by its adoption. Schumpeter outlines a series of roles displaced by capitalist political and economic frameworks: the old nobility, the peasant, the independent artisan, and more widely the social and support networks provided by the village and the craft guild. The gospel of disruption tends to limit itself to talking about the destruction of businesses, rather than of individuals or social classes, but there is no question that, in its recent application to the arena of education (Christensen and Eyring 2011; Christensen *et al.* 2011), it is aiming to undermine the social and humanist elements of traditional pedagogy by removing educators from classrooms, replacing personal teacher–student relationships with modularized and mass-produced online pedagogical material, and attacking the social structure of the teaching profession. Again, what we are looking at is a complex, self-sustaining social and economic network with real humans at the nodal points. This is what the status quo, so derided by contemporary advocates of innovation, consists of, and this is what is being protected when innovations are resisted and rejected. The disruptive and destructive tendencies of innovative technologies and practices impact these networks directly, subjecting some proportion of the people within them to increased harm, even as others might benefit. So, for example, to return to containerization, the commercial dock, when unpacked or unfolded, reveals a stunning complexity of networks encompassing the technological systems and engineers needed to produce each individual part; the collectively functioning parts that make up the boatyard, warehouses, and loading equipment; the dockworkers and their human relationships to each other, to the shipping companies that employ them (which are themselves black boxes containing multitudes), to their families, to the knowledge systems that allow them to do their various jobs, and to their union; the various packages of goods to be shipped, their individual contents, trajectories, patterns of ownership, and levels of insurance; the many ships that enter and leave port; and on and on, *ad infinitum*. These dockyards too are nodes in other networks that together form, for example, the global shipping industry or its constituent part – the multi-national transport company, itself a complex human-money-product-seascape-office-port hybrid. In introducing an innovation, in this instance containerization, this network-within-networks is the thing that is disrupted, and the flow of power, knowledge, and information through it must be redirected.

To borrow a metaphor from Marilyn Strathern (1996), innovations *cut* otherwise extensive networks. She uses examples drawn from the contemporary West and from Melanesian anthropology to suggest that specific interventions can stop the flow of information within networks, making their connections visible and cutting them at the same time. For example, she points to the transformation of the hepatitis C virus from a human blood sample; to a specimen tying together a complex scientific network; to a vaccine; to an invention to highlight the ways that an extensive and complex heterogeneous network can be rapidly and easily cut, in this case by the idea of ownership that adheres, in the form of a patent, to only six individuals of the myriad who studied and worked on the resultant vaccine. In cutting the scientific network, the various contributions and attachments of different scientists, labs, specimens, etc. are made visible, and only some are deemed worthy of owning the resulting vaccine. Moreover, this cutting process also brings into sharp focus the power imbalances that exist between nodes in the network, and has the effect of magnifying and reifying them – in this case, by institutionalizing a hierarchy of ownership over the hepatitis C vaccine. This lionized the contributions of six individual scientists and disregarded the rest. Using this formulation we can perhaps re-frame innovation resistance and conservatism as an effort to maintain the integrity of a network in which these power imbalances, while present, remain very much in flux and manipulable.

To put this in plainer language, although social and technological innovations are often treated as quite separate things (e.g., process versus product innovations), in actual fact technology and society are inextricably enmeshed. Changing technologies both provoke and rely on changing social mores and practices, and changing practices often involve new or different technologies. So to the surprise of few archaeologists, changing social conformations among Cornish people faced with an invading Roman army and their co-option of trade and mineral exploitation, changing forms of appropriate settlement, and the introduction of new ceramic forms and food styles must all be treated as related. Thus, resistance to a new technology may reflect a disinclination for the social changes which surround it – for the inevitable re-balancing of power relations and re-ordering of relationships, a process Goulet and Vinck (2017, 106) refer to as "detachment" – as much as a distaste for the new tool on offer (Bauer 1995, 19). For example, we can see in the case of the dockworkers' resistance to containerization (and likely also the Iowa farmers' resistance to new crop varieties) that the technology itself is not the focus of resistance; instead it is the social practices and networks of relationships (e.g., between individual people, between people and the wider community or specific social groups, and between people and things) with which it is entangled – both those that would allow it to flourish (i.e., shifting social networks, so that farmers spend more time in urban and educational environments interacting with non-farmers) and those that it would entail (e.g., a less-skilled, lower-remunerated, and smaller workforce on the docks).

In the case of the dockworkers, the role of their union in resisting innovation is crucial to highlight, as this makes clear the particular questions of power imbalance at the heart of this particular struggle. Unions are typically cast as the enemies of innovation in both popular tech writing (e.g., Ferenstein 2013) and academic management research (Bradley *et al.* 2017; Hirsch and Link 1987). This research argues that unionization has a statistically significant negative impact on firms' investment in product innovations and output, in the form of patents, but fails to address the negative impacts of decreasing unionization on the people employed by these firms. It is well known that income inequality has skyrocketed in the United States since the 1970s, and that, for most workers (especially those working for the minimum wage), wages have stagnated or even decreased. Economists and historians increasingly link these negative social outcomes to decreasing unionization, alongside the impacts of shifting immigration patterns and globalization (Gordon 2017, 613–18). Moreover, close studies of how unions engage with innovation have found a much more ambiguous picture, in which protectionism and apparent technophobia are mixed with careful balancing of new technologies with employee needs (see, e.g., Dubb 1999, Chapter 5). Again, what we see in these examples is a considered resistance to the social and political implications that form part of technological change, rather than a particular disinclination to innovate in a broader sense.

Indeed, new technologies can also be intrinsically socially conservative in ways that make innovation adoption less appealing. Although we often equate change with progress, itself a specious concept, change can be socially regressive even as it pushes technological envelopes.[3] The example of Uber, probably the best known and most divisive of the twenty-first-century Silicon Valley start-ups, is illustrative of this case. Uber's core innovations are (1) operationalizing the so-called "gig economy" to replace taxi services with an army of part-time drivers who use their own vehicles to take clients from place to place, and (2) connecting these drivers with potential passengers via a mobile phone app that allows pre-payment for rides and gives both driver and rider information about the other. While the mobile phone app is undeniably a new technology that has rapidly spread throughout the livery industry, the system of independent drivers using their own vehicles to service customers is much older. In fact, Uber's ride-sharing model looks surprisingly like the introduction, organization, and operation of London's nineteenth-century horse-drawn cabs.

In his unpublished Ph.D. thesis, Fu-Chia Chen (2013, 25–8) explains that the two-wheeled cabriolets ("cabs") were introduced in the early nineteenth century with more flexible, and eventually unlimited, numbers of licenses, to supplement (or perhaps more accurately, undercut) the heavily regulated and limited numbers of four-wheeled hackney carriages that had been the sole operators of horse-drawn livery in London since the late seventeenth century. Moreover, the typical Victorian cab driver was an "owner-driver," who owned a license to operate and two or three horses, as well as owning or hiring a cab. These cab drivers (like many of today's Uber drivers) were personally liable for a variety of road taxes and tariffs, and worked extremely long hours in order to pay these external costs, as well as to maintain their cab and horses and still have some money left to live on (Chen 2013, 35–40). Damage to the cab, or injury or illness to the horses, could prove disastrous, as there was little in the way of a safety net for these independent owner-drivers. The fare structure, while ostensibly regulated, was in practice largely a grey area, with cab drivers and passengers regularly at odds over the distance or time traveled and whether the rate agreed on for this was fairly calculated. Cab drivers frequently added an illegal surcharge (prescient of Uber's surge pricing) for inclement weather, or on holidays, such as Christmas (Chen 2013, 188–96). Drivers' associations, which offer some degree of security to cab drivers and license holders, emerged from a series of cab strikes over low rates and poor conditions in the nineteenth century alongside the framework of increasingly stringent legal regulation developed to ensure safe and fair conditions for cab riders. These are, perhaps ironically, the institutions Uber is now attempting to innovate away. Thus, it is no surprise that unions and drivers' associations, not to mention regulatory bodies, have resisted the dubious promise of Uber's ride-sharing model, since, in this case at least, the new technology is tethered to an unrepentant and deeply socially conservative return to the 1840s.

Resistance and persistence

The relationship between innovation resistance and the negotiation of power in society cannot be ignored. Bauer (1995, 14), in an important re-evaluation of the function and process of innovation resistance, suggests that resistance is better conceived of as an opposition to norms and the product of normative communication by institutions. Resistance, according to Bauer's formulation, breaks down norms, but this implicitly adopts the stand-point of the institution or power that is resisted. Alternatively, if resistance is protectivist in nature, such as the Luddite uprising carried out by workers who protested against the mechanization of the textile industry, it can be seen as supporting and reinforcing norms – in this case specialist craft industries – against transgression. In fact, Adrian Randall's (1995) re-interpretation of the Luddite movement, and more generally of the anti-industrial protests of the early nineteenth century, re-frames the strong and violent upper-class British reaction to the Luddites as effectively an act of class war against an increasingly threatening under-class. He suggests that the British upper classes perceived a direct link between machine breaking and the disruption, damage, and violence of the French Revolution, leading to swift and harsh reprisals for these acts of resistance to industrialization. By contrast, mechanization was slower in France, with the traditional craft guilds and craftsmen accorded more power and respect in the larger dialogues around industrialization. Randall implies (*contra* Mokyr 1990) that recent experience of the power of the French peasantry may have led to greater respect and success for their resistance to mechanization. Thus, in evaluating resistance to innovations we must ask who has the authority to define what is or is not valued, what is or is not traditional, and what is or is not progressive or beneficial – and for whom.

Anthropologists have regularly explored this interplay of power and resistance in the various ways colonized and Indigenous peoples have reacted to the Imposition of European practices, ideas, and technologies during the long history of colonialism in the world beyond Europe. Despite the repeated narrative of superior European tools and practices replacing Indigenous ones wholly and rapidly, Indigenous peoples regularly rejected and resisted European goods and ideas. They did so in complex and culturally specific ways, sometimes blending these outside innovations with traditional practices or technologies, sometimes taking one element of an introduced package and re-interpreting it entirely, or making strategic use of resistance to European innovations to push back against elements of the cultural and economic depredations of colonialism (Frieman and May 2019).

Chris Gosden (2004, 98–102) offers several examples of the way Oceanic people strategically rejected European goods and practices in order to retain traditional practices and power structures. For example, people from Tanga, New Ireland, made the decision to exclude European products from the competitive exchange networks upon which traditional social and political hierarchies were built, since they felt that the foreign materials were destabilizing traditional power structures. In much the same way, people in New Britain also chose to limit the penetration of European goods into ritualized exchange activities, even though these were used daily in non-ritual contexts (Gosden and Knowles 2001). In the Pacific, we can in fact see the development of a sort of doctrine of resistance in the post-colonial era in the form of *Kastom*. *Kastom* comprises a series of rules, practices, and beliefs grounded in pre-European Pacific identities and ways of life but reacting to and hybridizing with some European introductions. Its re-invention of "traditional" Pacific identities and re-interpretation of Pacific practices form key elements in the ongoing resistance to European religious and political structures (Keesing 1982, 1984). This is a particularly ambiguous form of resistance, in that it harnesses innovation to protect tradition rather than being simply conservative and anti-innovation. Specifically, when we look at *Kastom* and the practices associated with it, we are looking at a social framework developed by knowledgeable social actors who are carefully manipulating the social, technological, and economic framework of their world with clear aims – they are seeking not to return to an imagined unaltered pre-European Pacific culture, but to adapt Pacific values to respond to the challenges of the present and to resist their obliteration by globalizing colonialism (Lepowsky 1991).

Tradition is often framed as something static that stands in the way of – or in the face of – progress, in the form of new practices, technologies, or ways of life. To borrow another turn of phrase from Strathern (2004, 85), the idea of tradition and the idea of innovation both enfold many complex concepts. Yet, in the English language at least, they do not adhere to each other. By contrast, and building on decades of anthropological and literary research, Handler and Linnekin (1984, 287) describe tradition itself as a process, and one "that involves continual re-creation." Among archaeologists, Robin Osborne (2008) follows this framework and critiques the idea that tradition is merely a structure, a part of the habitus, that allows for the reproduction of society. Instead, he argues that social agency, in maintaining, translating, and transmitting traditions, must be at the foreground. Tradition, then, is both a frame through which, and a tool with which, to position social acts in dialogue with the past, with the wider context in which one is acting; with other people (both those who are members of the actor's community and those who are not); as well as with the future.

This complex view of tradition is in line with Indigenous perspectives on how we should understand innovation and resistance in the colonial process. Building on Native American critiques, Lee Panich (2013, 2020) suggests that decolonizing our ideas about innovation and conservatism means re-framing narratives of contact and colonialism as *persistence* narratives. In other words, by focusing on European innovations and how Native Americans responded to them, our research implicitly equates change with loss of culture. Instead, he argues that we should look at the vibrancy and flexibility of Native American traditions and their ability to encompass new materials, technologies, ways of life, and power dynamics that enabled Native Americans to persist and continue to develop a vibrant and living culture. In other words, tradition, and the acts of resistance that underpin its maintenance in this context, are only conservative from the colonizer's perspective, whereas, taking a Native American (or indeed, other Indigenous or Alter) stand-point, tradition is a dynamic, malleable, and highly reactive process in its own right.

Conservatism as a dynamic response to change

Although conservatism, resistance, non-, and anti-innovation are typically presented as the opposite of innovation, as we have seen they are themselves highly dynamic, creative, and thoroughly innovative processes in their own right. Thomas *et al.* (2017, 198) make this explicit, arguing that innovation is inbuilt within any act of resistance, which they define as "a form of critical innovation in which actors deploy strategies and actions of resignification, repurposing (re)design, production, implementation and management of technologies" in order to effect change. Even beyond the highly fraught power struggles of hegemonic dominance and colonialism to which Thomas and colleagues are responding, resistance and anti-innovative activities can be understood as playing dynamic roles within the process of innovation adoption. Bauer (1995) suggests we could better understand resistance to innovations as an active means of pushing against binary either/or decisions. Specifically, he argues that resistance functions to open or re-open debate – it slows or prevents the processes of stabilization and closure, and provides alternatives and options to apparently successful innovations or trajectories of social and technological development.

As an example of this we can look to Bruland's (1995, 128–33) narrative of the successful anti-innovation activism of the marginal fisher-farmers living in the Lofoten Islands and among the small towns and hamlets of Norway's Atlantic coast. For centuries, these communities opposed the introduction of a series of technological, political, and social innovations that collectively would have facilitated fishing in deeper waters and for larger catches. Instead, they fought to maintain the primacy of hand-line fishing with small boats in inshore waters. Over succeeding generations, their campaigns to protect traditional subsistence fishing practices were largely successful, with legislation passed that banned multi-line fishing in inland waters and enacted temporary whaling bans. In fact, continued tensions between low-productivity and high-productivity fishers fed into debates about European Union membership in the late twentieth century. Thus, we are able to see that a long-term pattern of resistance and anti-innovation activism has not only had an effect on the practice of fishing in Norway, but has also continually inserted itself into wider (but related) social and political debates – forcing repeated negotiation and the development of alternative practices into technological systems, political relationships, and social structures while preventing rapid stabilization and closure.

This example is particularly potent from an archaeological perspective since, at the chronological scale visible to us, we must understand conservatism as both a constellation of small-scale social choices – a pattern of regular and repeated innovation resistance and rejection enacted by individuals – and the outcome of social practices that played out over generations. We can certainly observe a similar pattern in Iron Age and Romano-British Cornwall. Conservatism here emerges from and is enacted through the desire to curate and display heirloom objects as symbols of power; the choice to inhabit carefully maintained ancestral settlements; and, particularly, the continuation of local technologies linked to cultural transmission and to the highly personal and emotive domains of food preparation and consumption. Yet, Cornwall in this period was not unchanging – nor was it isolated from outside innovations. Indeed, many of the sites discussed above – Hall Rings, Bake Rings, Trethurgy – lay within an easy day's travel from at least one of the three Roman forts currently known in Cornwall. Instead, people were carefully manipulating the outside material and practices they adopted, as visible, for example, in the continued use of gabbroic pottery to make new forms of vessels for new styles of food consumption in the Romano-British period.

In the contemporary world, we can observe this dynamic in practice among communities who intentionally limit their contact with modern technologies and society, such as the North American Amish. Whilst popular portrayals of these communities focus on their old-fashioned attire, horse-drawn vehicles, and perceived opposition to "technological progress," actually these are the result of a series of ongoing negotiations about social values, community connectedness, and the maintenance of tradition. Work by Jameson M. Wetmore (2007) based on research among the Pennsylvania Amish makes clear that these communities are very aware of the scope and functionality of the technologies and practices they eschew, but prioritize tradition and "fellowship" over machinery. They are cognizant of the shifting relationships and power structures that accompany technological change, so only those technologies that they do not see as disruptive to their social structure are deemed licit. Moreover, just as we have seen above, in the case of Amish reflections on technology and society, tradition and innovation are not in contention but are enfolded within each other – the need for technologies that both enhance and respect tradition result in extremely creative and surprisingly inventive practices. For example, Wetmore points to the development of non-electric versions of acceptable electric technologies, including carpentry tools and refrigerators retrofitted to work on battery power, diesel generators, or pneumatic pressure.

Lindsay Ems (2014, 2015) has observed exactly the same careful negotiation of power, technology, and practice to exist in other American Amish communities. For example, she observed that Amish workers who contracted out for labor found themselves in need of a telephone, both to liaise with clients and to communicate with each other from far-flung worksites, but mobile phones were considered to be too disruptive. Instead land-line phones were retrofitted with attachments to interface with the cellular network and to be powered by a car cigarette lighter. These were placed in plywood boxes so that they could be transported out to worksites by construction crews but would remain inconvenient for personal use (Fig. 6.4). Instead, the one phone is shared by the whole work crew and remains with the vehicles and equipment – a highly innovative way of resisting the pressure to carry individual mobile phones.

By now, it should be clear that I do not view conservatism (and the many innovation-avoidance and resistance practices enfolded within it) as necessarily separate from or antithetical to innovation. In fact, acts of resistance, or the re-invigoration or maintenance of tradition, typically entail a deeply creative praxis, and include and enfold myriad

Figure 6.4   An Amish "black box" mobile phone.

innovations in their own right. Moreover, innovation and change can also be deeply conservative in nature, hearkening back to earlier practices or outdated technologies. In order to understand conservatism, we need to remove innovation from the question entirely. With all its contemporary valuation and implications, it is a distraction from the choices people made, are making, and will make, and from the long-term trends that together give us the impression of a group of people who are less likely immediately to adopt a given innovation or who are willing to resist, actively and passively, in order to maintain their social, political, and technological network.

In the 1970s, Allan Mazur (1975) attempted to develop a logical explanation for the opposition to technological innovation. He studied social movements against fluoridation and nuclear energy, and came to the conclusion that most opposition stems from a perception of risk – he focused on physical dangers, but the danger to social structures is certainly also relevant – and how these are communicated via social influence. In other words, the influence of friends, family, and trusted community members was key to convincing individuals that a risk was present and should be resisted. Just like the early adopters or community leaders treated as key vectors for the spread of innovations in diffusion models, these community leaders diffused resistance. Moreover, he observed that not only were the leaders of opposition movements well-integrated members of society, they spread their message not individually but via already-organized networks – school groups, church groups, etc. This is certainly the case among the Amish, but is also true of the Lofoten fishermen. Bruland (1995, 133) notes that the national politicians who voted in legislation to protect traditional fisheries and fishing practices in the nineteenth and early twentieth centuries, far from being a distant urban elite, were in fact from impacted villages, and maintained close social and economic ties to these communities. Like the Cornish rural heterarchy, their leadership structure was relatively flat and power was distributed. In other words, the very same constellations of communities of practice discussed as key vectors for the diffusion of innovations in the previous chapter play entirely the same role in maintaining traditions and dispersing resistance.

Organizing effective resistance movements, especially against a strong power, requires trust and solid communication networks that can operate around authorized channels – submerged networks by necessity. The act of resisting an innovation can further enhance the closeness of these ties. For example, in the case of the development of *Kastom* in the Pacific, we see new post-colonial Pacific identities being created in opposition to European culture, practices, and technologies that are then strengthened through their ties to local traditions and re-invented practices. As Randall

(1995, 64) notes in the context of the English Industrial Revolution, communities with "a long history of riotous protest" were the most likely to engage in machine breaking and to resist the industrialzsation of their crafts, not just because of the social cohesion of an already-organized community network but also because community leaders emerged who were knowledgeable in and practiced with successful tactics of resistance.

So, it seems, we must re-frame conservatism away from a trait that characterizes the reluctant, risk-averse, and backward. Those who shy away from one or many innovations are not less educated or less progressive, but instead are building on what has been a successful and likely long-term social dynamic within their communities. That is not to say that these communities are less creative or inventive than their neighbors who appear more open to innovations. As we have seen, tradition is innately innovative, and resistance – particularly in the face of dominant cultural and technological norms – requires flexibility and creativity in order for socially acceptable workarounds, justifications, solutions, and strategies to develop. Conservatism, then, is not only not monolithic, it is markedly innovative and highly dynamic in its own right.

Notes

1 This is not to say that the "old man shouting at clouds" stereotype has not been applied to social conservatives but, in this case, it is largely a meme circulated among social innovators rather than a widespread stereotype. If social conservatives were genuinely widely understood to be backward-looking and ridiculous, the United States would not have elected Donald Trump president, Brexit would not be the law of the land in the United Kingdom, and Australia's Parliament would have legalized gay marriage without blinking, just for example.
2 I've borrowed this evocative turn of phrase from Twitter commentary by Dr. Sarah May, https://twitter.com/Sarah_May1/status/1255760843817525249 (accessed September 17, 2020).
3 Indeed, the opposite is also true, and is likely the source of many of the anti-union myths about technophobia touched on above. Great social change – for example, a shift to a well-remunerated blue-collar workforce – does not require new technologies and, in fact, could be argued to be maintained more easily in a less automated world.

7

Create/innovate

What leads to innovativeness? Are some groups really more creative than others?

If previous chapters of this book concerned themselves with asking what innovation is and how it operates as a social practice, then perhaps this chapter is best read as my attempt to answer the rather abstract question "Why is innovation?" In the previous chapter, I explored conservatism, which I pulled apart into various threads – tradition, resistance, continuity, persistence. In this one, I conduct the same sort of dissection of innovativeness. I have already suggested that most archaeological discourse surrounding the adoption of innovations has been hampered by a reductive and functionalist "Do-Need" framework. That is, that innovations are typically assumed to be adopted because they *do* something that fulfils a *need*. In this model, innovation serves to address either external "push" factors or internal "pull" factors. In other words, these models position innovativeness as an innate reflex that kicks in when the right stimulus is applied. Yet we have already seen that all of the elements that lead to the adoption or rejection of an innovation are deeply socially, spatially, and temporally contingent. The innovation process is messy, non-linear, and contested, so I argue innovativeness must be as well.

Innovativeness is perhaps best understood as the tendency – personal or societal – to embrace innovation; to experiment or invent; and to adopt new things, techniques, and ways of life. Indeed, this definition itself raises a primary problem for our understanding of how innovativeness appears or operates: our attention is immediately split between individual traits and societal patterns. Archaeologists, of course, have little to no access to the self-perception and identity formation of individuals in the past, let alone their ideas about innovation. We often are constrained in our approach, forced by the limitations of our data to gloss over the individual in our work. Ours is largely a normative discipline, one that constructs its narratives from the gross commonalities of the fragmented material record and seeks to build patterns of common behavior from these. However, the turn to agent-centered approaches in recent decades offers an alternative perspective. Agency theory, which emerges from foundational anthropological and sociological work by Bourdieu, Giddens, and colleagues, takes the position that individual members of society can creatively manipulate the larger social structures that frame and order their relationships, their practices, and their embodied experiences (Dobres 2000; Dornan 2002). Archaeologists working within this framework have re-oriented attention onto non-normative features, activities, or traces in the archaeological record – effectively those elements that might be read as showing evidence of creativity or creative practice. Following this logic, I would argue that non-normative behavior gives us insight into both scales of innovativeness: that of the creative individual as well as that of the broader social norms or structures that frame and orient (but do not necessarily constrain) their behavior.

For the purposes of this chapter, I will more or less refer to societal-scale innovative behavior, structures, and practices as *innovativeness*, and to the personal-scale as *creativity*. I recognize that this is an imperfect division – and one that is not entirely in line with scholarship on these topics in other disciplines, notably psychology and organizational studies. However, I am employing this dichotomous approach specifically as a heuristic to allow me to clarify an argument made at two scales about similar, but not quite identical, topics. In this chapter, I draw out a comparison of these two intertwined phenomena – creativity and innovativeness – in order to examine the social factors that might affect how we recognize and perceive them in the past.

As discussed in previous chapters, colonialist and Eurocentric ideas about technology, social dynamism, and progress have deeply colored (and, as we shall see, continue to color) our ability to discern and understand cross-cultural differences in the expression and valuation of creative practices and innovative behavior. Here, I attempt to plot a course through this minefield in order to discuss not just whether some people were or are more innovative or creative than others, but how we archaeologists can assess apparent differences in these areas. I begin by exploring how creativity and innovativeness have been identified and perceived among our hominin ancestors. The origin of creativity has long been sought by Palaeolithic archaeologists, and its distinctiveness (or not) to our own species is still debated.

Human origins and the origins of creativity

To be creative, to create and innovate, is apparently unique to humans, but whether it is a distinguishing feature of our specific species or part of the broader spectrum of hominin behaviors is not agreed upon. Indeed, the ability to debate this point is rendered more difficult not only because of the limitations of the archaeological record from periods beyond hundreds of thousands of years old, but also by a discourse around creativity and humanity that emerged in the nineteenth century and positioned modern humans – *Homo sapiens*: the *wise* humans – as uniquely innovative among the members of our genus. Art, music, language, and social organization were all associated with *H. sapiens* and presented in contrast to the obviously disconcerting Neanderthals. Over a series of publications in the 1990s, Stephanie Moser illustrated how depictions of Neanderthals and other non-human hominins followed European racial and gendered stereotypes (Moser 1992, 1993, 1998). In other words, the colonial lens that perceived western Eurasian people as more civilized than the non-white peoples whose land they settled was turned on newly discovered hominins, who were then given all the same stereotypical traits believed to characterize non-European peoples. This, of course, included a lack of creativity (Zilhão 2012).

We can see the fallacy of these stereotypes and misconceptions by comparing nineteenth-century European perceptions with contemporary representations of Indigenous peoples from around the world – most notably Aboriginal Tasmanian people (discussed in Chapter 1), whose apparent (if false) lack of innovation (and, thus, creativity and, thus, humanity) contributed to the remorseless genocide carried out by British colonists. Indeed, early classifications of our species separated out different geographic populations of humans and attributed them to different sub-species – in Linnaeus's classification, for example, we see *Homo sapiens europaeus*, *H. s. afer*, *H. s. asiaticus*, and *H. s. americanus* (Linné 1758). Increasing numbers of sub-species were proposed in subsequent generations, and the explicitly racist agenda of this project can be seen in the evolution of *H. s. europaeus*, who first became *H. s. albus* (that is, the wise white man) (Linné 1788) and, in the 1940s, was distinguished as *H. s. sapiens* (the wise wise man) even while other sub-species retained and were qualified with geographical or racial descriptors. E. Raymond Hall (1946, 359), for example, distinguishes five sub-species of *H. sapiens* and identifies their representatives among living populations, which are described in racial rather than geographic terms, so *H. s. sapiens* are "Caucasian" while *H. s. americanus* are "American Indian" and *H. s. tasmanianus* are "Australian Black."

In fact, while most of these sub-species identifications were discarded in the final quarter of the twentieth century,[1] *H. s. sapiens* – our doubly wise selves – has remained in use in places (e.g., in popular communication, by the Australian Museum, as well as in scholarly contexts, such as Mithen 1998, 175).[2] Modern usage, however, sees the sub-species description broadened to include all modern humans (not just white Europeans) in order to distinguish the humans of the last hundred-or-so thousand years from archaic *H. sapiens*, who evolved in Africa sometime around 300,000 years ago. This distinction is made on the grounds of perceived differences in behaviour and cognition, rather than biology. Mithen (1996, 1998), for example, builds on Pfeiffer's (1982) earlier work to suggest that cultural and cognitive developments took place between 100,000 and 30,000 years ago that led to the emergence of modern human practices (see also Klein 2000). Paul Mellars (1996) suggests eight specific behaviors (or at least the material traces of these) that can be linked to such a cognitive revolution: the emergence of blade (rather than flake) technology in lithic production; a sudden increase in the types of stone tools; the emergence of bone, antler, and ivory tools; an increase in the pace of technological change; the appearance of beads and other bodily adornments; the invention of musical instruments; the first representational art; and shifting economic and settlement practices. Mithen's influential interpretation of these data is that this "creative explosion" is due to the emergence of cognitive fluidity in modern humans: that is, the ability to cross-pollinate domains of knowledge (Fig. 7.1).

However, in the last few years, the idea of drawing a hard (or even dashed) line between archaic and modern *H. sapiens* has been questioned, in large part because of our increasing knowledge of the social practices of non-human hominins (Nowell 2010), as well as an increasingly rich archaeological record that pushes back the emergence of art, adornment, etc. much earlier than previously thought (d'Errico and Stringer 2011; Gamble *et al.* 2011; McBrearty and Brooks 2000; Wilkins 2018, 2020). Only twenty years ago at time of writing, Kuhn and Stiner (1998) struggled with an apparent disjunction of the evident ability of Neanderthals to adapt creatively to new environments and raw materials with the apparently nearly static Mousterian lithic technological tradition, which saw few innovations in the form of artefacts produced. They characterized this lack of technological diversity as "strange," and identified it as a central quandary in their attempts to understand how or if *H. neanderthalensis* could be conceived of as creative.

Indeed, as their struggle makes clear, defining creativity for these early periods is often as much a problem as assessing how creative ancient hominins may actually have been. This is a particular conundrum for archaeologists, since our discipline's underlying intellectual framework is strongly Eurocentric (McBrearty and Brooks 2000, 534), meaning that European (and often purely Anglo) conceptions of creativity dominate, even though we know from research in the wider human sciences that creative expression is culturally contingent (Morris

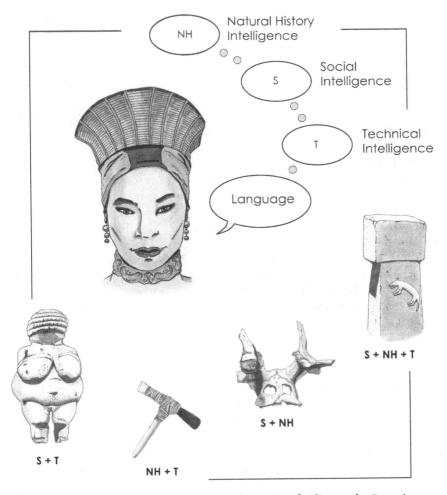

Figure 7.1 Mithen's model of the impact of cognitive fluidity on the "creative explosion".

and Leung 2010; Rudowicz 2003). That is, our dominant models associate creativity with novelty, especially as regards technological innovation (of the sexy product-innovation variety), and do not give equal weight to other forms of innovativeness, such as process innovation, incremental developments, or socially acceptable (as opposed to highly original) solutions. Shifting our focus away from the production of stone tools, we can observe that not only were Neanderthals able to adapt their lifestyles, diets, and practices to a range of environmental contexts – some quite extreme – but they also made use of colorants, likely for bodily ornamentation alongside perforated shells and teeth (Zilhão 2012).

Along these lines, Mark Lake (1998, 125) has suggested that the entire *Homo* genus might be conceived of as *H. creatrix*, with all related species distinguished from other primates by their ability and capacity for creative problem solving and technological innovation. While primates (and other species) have been documented as using tools, he argues that only members of the genus *Homo* have demonstrated a capacity for imagination and forethought – visible in their choice of dwelling and food-gathering locales; their ability to imagine and then reproduce the spatial properties of stone (and presumably other organic material) tools; and their capacity for social learning, as demonstrated in the transmission of these forms and the special techniques by which they were produced (see also Gamble 2012; Gamble *et al.* 2011). In his recent book *The Smart Neanderthal*, Clive Finlayson (2019) builds on these observations through the lens of Neanderthal relationships with birds. He carefully and elegantly makes the case not only that Neanderthals hunted birds for their flesh and feathers (*contra* decades of earlier received wisdom), but that their ability to plan around seasonal resources and the movements of animals – as exemplified by migratory and sometimes aggressive avian species – proves their cleverness and creativity. He sets this alongside recently published examples of Neanderthal rock engravings (Rodríguez-Vidal *et al.* 2014), cave paintings (Hoffmann *et al.* 2018b), and symbolic use of mineral pigments (Hoffmann *et al.* 2018a) to argue strongly against the persistent misconception that *H. sapiens* is a uniquely wise hominin.

Although it becomes harder to discern the farther back one goes, there is abundant evidence that other hominins were also creative – if not, perhaps, exhibiting the same cognitive adaptations as us or our Neanderthal cousins. According to current models, *H. erectus* (following current scholarship, an ancestor of both the human and Neanderthal lineages) began a slow migration out of Africa some 1.9 million years ago, spreading first eastwards to what is now southeast Asia and Indonesia, and later westwards into southern and southwest Europe. Carotenuto *et al.* (2016) point out that, across Eurasia – and quite distinct from Erectus behavior in Africa prior to their wider dispersal – they seem to have preferentially occupied areas with copious, easy-to-collect, and ready-to-use stone for toolmaking. However, recent research into Erectus' stone procurement and usage strategies in the Arabian Peninsula paints a mixed picture of their innovativeness. Shipton *et al.* (2018) present a carefully excavated Acheulean assemblage from Saffaqah, Saudi Arabia, dated to at least 200,000 years ago. They suggest that the hominins who produced this assemblage were dexterous, technologically sophisticated, and knowledgeable about the resources available to them, but not creative in their toolmaking, since there is no change in technology apparent from the oldest to most recent layers at Saffaqah. Certainly, local resources were preferred, but in Java, between 540,000 and 430,000 years ago, for example, this meant that Erectus developed cutting tools of shell in place of stone. Here we also see the likely use of shark teeth for opening bivalves and perhaps also for engraving them (Joordens *et al.* 2014).

Beyond tools, Erectus seems to have been the first hominin species to control and use fire from at least 1 million years ago (Gowlett 2016), and more consistently from about 400,000 years ago. Controlling fire implies both a technological innovation, but also a creative way of engaging with the wider environment – fire could be (and likely was) collected and maintained as embers from natural blazes, just as it could be kindled. Control of fire also implies behavioral changes, including changing dietary practices. Fire could protect a group from predators. It could illuminate the night, extending the time people were able to socialize and work together. It could be used in hunting and to make tools. Indeed, there is a good case to be made that the impetus to make tools at all – a trait shared by many contemporary non-human primates, as well as our most archaic relatives *H. habilis* and *H. naledi*, in addition to members of the *Australopithecus* and *Paranthropus* genera – also represents a creative manipulation of the natural environment, despite differences in cognitive ability or brain size (Hovers 2012).

While many species migrate as climatic conditions change, hominins have proven uniquely able to live in naturally hostile environments for which we were poorly adapted. Tools – made of stone, shell, leather – as well as more broadly conceived technologies – such as pyrotechnology or cooking – and perhaps also social practices, such as the communal rearing and protection of vulnerable children, and complex social practices that allowed for processes of teaching and learning, likely contributed to our ancestors' ability to thrive and survive. Stated baldly, it seems obvious that Erectus and Neanderthals were (like us) creative problem solvers, able to innovate in order to deal with changing environments and newly introduced ideas, as well as creative in their personal expression and social relations. Nevertheless, the idea that we *H. sapiens* are uniquely gifted with creativity persists in both scholarly and popular literature (e.g., Harari 2014). Finlayson (2019, 196–7) calls this a fiction, residual of an earlier age; but fiction can be dangerous. Following Porr and Matthews' (2017, 2020) recent critique, ascribing a tangible and biologically significant distinction between early and recent *H. sapiens* (not to mention between recent *H. sapiens* populations and other, earlier hominins) on the basis of behavior seen as more sophisticated, more cultured, or more modern than the first members of our species to emerge is a racist fallacy based on the same colonialist logic that deemed non-European sub-human. The lesson here is that we must be careful in assessing what exactly we assume to be creative behavior, lest we replicate reductive and essentialist distinctions of racial difference.

Archaeology and creativity

Creativity is hard to grasp in an archaeological context. It is, according to the majority of psychologists who have studied it, a product of divergent thinking – a deeply interior phenomenon in which the form and generation of a person's thoughts are allowed to stray from normative and linear patterns. It is thought both to sprout naturally from innate tendencies towards non-conformity and risk-taking, and to be germinated, cultivated, and perhaps even forced with proper fertilization and pruning. I use these gardening metaphors intentionally, because creativity is often likened to flowers: blooming in bright and unexpected colors, an aesthetic and unalloyed positive addition to any environment. Although many different behaviors, from play to art making to technological invention, are lumped together in the creativity-studies literature,[3] Margaret Boden (2004, 1) offers us a quite reasonable definition, writing that "creativity is the ability to come up with ideas or artefacts that are *new, surprising and valuable*" (italics in original). That is, creativity implies behavior centered around the production of a *thing* (tangible or intangible) that is both novel and useful (functionally or socially or both).

Following this broad definition, creativity – or at least the material traces of it – should be easily observed by archaeologists. We find *things* all the time and, through careful comparison and even more careful dating regimes, we are often able to identify those that are newer than the rest, as well as to make tentative feints towards understanding their value and function. And, yet, isolating the creativity – whether in form or technique – from an endless sequence of new things is not straightforward. Where archaeologists approach creativity, they are typically looking at novelty in form, technique, or function (as do Leusch *et al.* 2015) or artistry, that is, the emergence of new types or styles of art making (e.g., Malone and Stoddart 1998). These approaches are largely descriptive, outlining what the new thing is and when it emerged, sometimes framed by a discussion of new resources or environmental constraints. Cognitive archaeologists have explored the neurological or evolutionary roots of creative behavior during the evolution of hominins and human society in deep prehistory (Mithen 1996; Renfrew 2009), but with contradictory results, as discussed above. Indeed, art making and artistry loom large within their work as well, being among the key creative outputs used to distinguish so-called behaviorally modern humans from other hominins (or even other populations of *H. sapiens* who did not make art, or whose art has not survived or been recognized as such) (Bar-Yosef 2002).

Certainly, the wider field of creativity studies is itself contradictory with regard to what creativity is, how it is expressed, and whether it is irregularly distributed among the human population. Although we still point to J. P. Guilford's (1950) mid-twentieth-century call to study creativity and to celebrate the creative genius, scores of research projects have since been established that underscore that there is, in fact, no such thing as a creative genius. Instead, while some individuals are more prone to risk taking and divergent thought, creativity and its expression are perhaps better understood as existing along a spectrum (Liep 2001, 6). Moreover, the identification and expression of creative thought and practice are strongly colored by social context.

All people are creative, though creativity might be expressed differently in different cultural settings (Erez and Nouri 2010; Grigoryan *et al.* 2018; Morris and Leung 2010). Societies that place a higher value on authoritarian principles, traditional practices, and collective decision making may be more hostile to creative practice and conceptions of creativity that lionize non-conformity, singular creative individuals, and abrupt and large-scale change (Ng 2001). However, individual members of these societies are no less creative than people growing up in individualistic and change-oriented cultures (Lubart 1998, 2010). In his influential psychological model of creativity and innovativeness, Kirton (1976, 622) posits that inventiveness and adaptiveness – that is, the creative impulses to "do things differently" versus to "do things better" – are two poles of a spectrum of personality traits shared cross-culturally, but that people falling at the extreme ends of this spectrum might find themselves rather isolated from their peers. Creativity, then, is universally present but shaped by society, and expressed in ways that emerge from and respond to both tradition and new social, environmental, and technological realities.

With this in mind, it is worth exploring the work of Joanna Sofaer, one of the few archaeologists to have made a careful and wide-ranging study of creativity in archaeology (Jørgensen *et al.* 2018; Sofaer 2015, 2018). Instead of falling into the trap (discussed in Chapter 3) of looking for (non-existent) solitary geniuses, she re-frames creativity as a social practice that

> emerges within social settings in which knowledge is cultivated and transferred among people. It can be understood as the ability to see connections and relationships where others have not ... In other words,

> innovation in the production of cultural forms involves the manipula-
> tion, reconfiguration and recategorisation of familiar forms and ideas ...
> Creativity is a matter of knowing how to work with [materials] – their
> potentials and limitations – in order to put ideas into practice. (Sofaer
> 2015, 2)

In other words, she argues, first, that we can understand creativity as an externalized social phenomenon, rather than an interior orientation of singular individuals and, second, that understanding creativity requires us to explore the embodied knowledge and practice of past people. In her recent monograph *Clay in the Age of Bronze*, she examines the expression of creativity in Bronze Age southeast Europe through various engagements with ceramic technology – the production of figurines, the use of pottery in funerary contexts, the failure of new vessel forms. She highlights the spectrum of creative practice, from everyday problem solving to bursts of novel (and often quite artistic) technological practice.

Of particular interest here, she emphasizes the links between pre-existing knowledge – of materials and their physical properties, of cultural practices, of neighboring communities – and novel expression. In her discussion of creativity and performance, she highlights the example of the dramatic funerary rites at the second-millennium-BCE cemetery of Cârna, Romania (Sofaer 2015, 137–48). At this site, cremation was the dominant rite, with the cremated remains of single individuals being deposited in decorated ceramic vessels that were then adorned or elaborated through stacking a number of other vessels on top of and around them. While the rite itself is relatively standardized, with some difference in number or type of vessel based on the deceased individual's age, gender, lineage, or social standing, its *performance* would have been full of tension and emotive expression. It required at least two (if not four or five) individuals to position and stack the pots, many of which seem to have been made for display within this type of rite, and failure (broken vessels, collapsed stacks) would have been an ever-present possibility. Moreover, since no two stacks are identical in form or content, even if the script of the funerary rite appears to have been more or less standardized, its embodied practice allowed for and promoted novelty and surprise. In this way, tradition and novelty were bound together in the Cârna funerary rite. Although the rhythm and beats of the rite were well established, the gestures and outcome would be unexpected. Following Sofaer, this dramatic tension allowed for the relationships between living and dead to be re-configured while the construction and eventual burial of the stack of pottery signaled the end of the funerary rite and transformed the deceased from viscerally and physically present to abstract memory.

Taking our cues from Sofaer, we can position creativity as a mediating force between tradition and the unknown future. In their critique of the creativity-studies discourse, Rehn and De Cock (2009) emphasize that the central focus on novelty and originality in this research is an ideological construct, embedding creativity in modernist progress narratives and devaluing histories of labor and collaboration. They argue that creative acts by necessity engage with tradition, even as they sometimes break with it. Indeed, this meshes well with the model of tradition developed in the previous chapter, which positions it as an active engagement with past practice and present realities that emphasizes links with the past while adapting to the context and constraints of the present. Tradition, then, and its constant re-creation, are creative acts.

Play and improvisation – two core elements of creative practice – both work within these traditional structures and riff off them, creating space for dissent, resistance, and subversion (Rosaldo *et al.* 1993). They pick at the edges of established behavior, belief, and organization, pulling loose threads and unraveling assumptions. Recalling our discussion of peripherality and innovation in Chapter 5, it is perhaps for this reason that marginal areas and interstitial spaces are so rich with creative potential (Dogan 1999). Sofaer (2015, 76–8) positions marginal creativity as a product of creolization or hybridity. We might, for instance, recall the discussion of Australian contact rock art and knapped glass from Chapter 4, and particularly Harrison's (2003) suggestion that knapping old lithic types for sale to Europeans was a form of playful subversion of colonial norms. The flurries of creative design that underlie these practices were possible, in part, because of the cultural and geographic remoteness between Aboriginal people and European colonizers. This sort of creativity made space for Aboriginal people to take ownership of and comment on invasive European things and ways of life. While direct, armed resistance to colonial incursion occurred (and met with extreme violence[4]), creative and playful activities opened avenues for subversion and social commentary that would have been largely invisible to colonial authorities, but readily visible to fellow Indigenous people (Harrison 2003).

Indeed, even beyond the disparities of power between settler and colonized populations, negotiating cultural and geographic remoteness fosters and supports creative expression and design (Gibson 2010). As Clay Shirky (2015) put it in his discussion of the vibrant maker community in contemporary China, "thin wallets are the mother of invention." In other words, a lack of access to materials or resources can spur and make acceptable creative workarounds, experimentation, and new forms of problem solving. Although geographical remoteness from resource-rich centers can pose challenges (particularly economic ones), it also provides freedom from the dominant "trends, fashions and compulsions" (Gibson *et al.* 2010, 31), as well as from the baked-in or path-dependent expectations of well-known processes and their presumed results. Indeed, the creative expression of marginalized communities – the queer community, ethnic minorities, etc. – often has an out-sized impact on popular culture: the tastes, dialect, music, and over-arching style of the margins can be and often are appropriated by the mainstream (e.g., Thomas 1995). It is powerful and attractive because it is subversive, clever, and – by its very nature as marginal – boundary-crossing. Following David Gooding (1990, 168), creative experimentation results from situations of uncertainty or marginality, since not knowing the solution in advance means that all possible solutions are equally feasible. So, while creativity as an internal and innate psychological feature is universal, different social, environmental, and economic contexts affect the scale of its expression and impact. Thus, it seems likely that the innovativeness we observe at a social level might reflect the conditions and relationships that enable or inhibit the expression of creativity.

Understanding innovativeness

Innovativeness, or the general propensity to innovate through invention or adoption, has been linked to numerous personal, psychological, political, and social attributes. As Grigoryan *et al.* (2018, 337) summarize, these include values, trust, patterns of education, knowledge transmission patterns and relationships, power differences, acceptance of individualism, and socio-political organization, all of which have been examined as factors in innovative behavior. Midgley and Dowling (1978), for example, critique earlier work (including Kirton's 1976 study cited above) that equated innovativeness with personality traits. They argue that innovative behavior is complex and strongly affected by personal relationships as well as by social and economic factors (Fig. 7.2). Elizabeth C. Hirschman (1980) built on this model in a highly influential paper to argue that innovativeness on the part of contemporary consumers could be broken into two distinct categories: adoptive innovativeness (openness to adopting new things) and use innovativeness (openness to re-purposing already-known things, or to inventing new uses for them). Both of these, she argued, are influenced by personality traits, but also by social factors, such as access to and interest in information about innovations (which she termed "vicarious innovativeness"), and the number and type of social roles played by any given consumer ("role accumulation"). However, much of this research – and the wider literature that cites it (e.g., Price and Ridgway 1983; Ram and Jung 1994; Ridgway and Price 1994) – is tightly focused on modeling

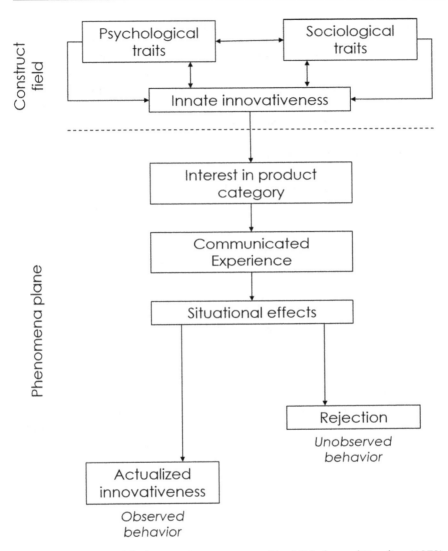

Figure 7.2   The model of innovativeness proposed by Midgeley and Dowling (1978).

the reactions of individual consumers, rather than looking for broader social trends.

As we saw in the previous section, population- or culture-scale discussions of creative or innovative behavior are rather fraught, drawing as they do on national stereotypes and – recalling the case studies of Indigenous innovation in earlier chapters – racist and colonialist social models. However, from a historical perspective, it seems evident that innovativeness might differ between groups as well as individuals; and innovations certainly seem to cluster in certain industries, sectors, nations, or communities, implying larger structural processes at work (Fagerberg 2006, 20). One attempt to answer this question was offered by the economist Scott Shane (1992), who developed a model of socially differentiated inventiveness through an examination of patent data from firms in thirty-three countries. Based on this analysis, he concluded that two of the key factors influencing innovativeness are relative power difference – that is, whether power flows vertically or horizontally – and the extent to which individuality is tolerated. Specifically, a more individualistic society or organization with a less hierarchical structure will be more innovative than one that places a higher value on conformity and in which power and authority are centralized. Although patenting rates do not make an obvious proxy to archaeological data, this finding certainly seems to imply that smaller-scale societies with fewer entrenched status differences – exactly the sort of society in which most people have lived since we evolved – should be prone to or at least more open to innovation. Certainly, this also applies to specific organizations, for example craft guilds, within more hierarchical social modalities.

Shane's observations compare favorably to research in other fields. Economic historian Joel Mokyr (1992, 328), for example, builds on his research into the British and French industrial revolutions to suggest that "centralized bureaucracies ... breed conformity," not creativity. Similarly, while Flemming Nielsen's (2001) study of farming innovations in Kenya made clear that all farmers innovate to some degree (though this is mostly minor and pragmatic, taking the form of Hirschman's "use innovativeness" – i.e., optimization or re-purposing of existing technologies or practices), he did observe small differences linked to gender, age, and ethnicity. He found that female-headed households, households headed by someone under 60 years of age, and ethnic Luhya farmers tended to be more innovative. He tentatively suggests that the latter might reflect a less hierarchical and age-stratified social organization among the Luhya than among neighboring (and less innovative) Luo farmers. He does not suggest reasons why age or gender could affect innovativeness, but we might speculate that younger farmers may be less risk-averse than the older generation, and that women running farms in a patriarchal society are also less likely to respect or pay significant heed to centralized authorities or traditional norms.

Despite this body of research demonstrating the complex social and political factors that influence cross-cultural innovativeness, much of the archaeological discourse around this topic still seems to default to the simple economic/environmental push–pull factors discussed above. Groups are innovative either because they are obliged to invent or adopt new technologies by external pressures – environmental, demographic, economic, etc. – or because they perceive a social or political benefit resulting. Thus, the adoption of agriculture has been modeled both as the direct result of climatic changes (Bar-Yosef 2011) and as a teleological process spearheaded by cunning, aggrandizing individuals, who saw it as a means to become and remain social and political elites (Hayden 2014).

This discourse is particularly apparent when we look at innovativeness and creativity during the Palaeolithic. Francesco d'Errico *et al.* (2018), for example, contextualize their discussion of the emergence of complex sewing technologies in Eurasia during the Upper Palaeolithic by positioning it against the expansion of *Homo sapiens* into cold regions and addressing their capacity to adapt to climate change. This frame is unsatisfactory because not only were our direct ancestors not the first hominins to live in these cold regions, the Upper Palaeolithic was certainly not the first period in which they had to react to climate change. Nevertheless, complex sewing kits have yet to be found from contexts dated earlier than around 45,000 years ago. Moreover, the authors themselves note that the earliest needles appear to have had specialist uses relating to ornamentation of clothing items, rather than the invention of thermal regulating attire, implying a rather less environmental story behind their development. However, studies of innovation and innovativeness in this remote period are immeasurably complicated by issues of preservation and the low precision of scientific dates. It is hard to assess a cultural – as opposed to species-level – propensity to innovate when our margins of error are thousands (if not tens of thousands) of years. Consequently, I want to turn briefly to a (much) more recent period in order to explore how innovativeness manifests as a complex social and personal phenomenon, even as it is influenced and shaped by constraints, such as environmental pressure.

Innovative problem solving in the Cornish mining diaspora

During the mid-nineteenth century, tens of thousands of skilled Cornish miners left home and traveled around the world to mine copper, silver, gold, and other metals in places as distant as California, Canada, Spain, Chile, Argentina, South Africa, and Australia. The Cornish diaspora represented just one thread in the complex mesh of emerging globalization within the British Empire. These meandering miners retained a strong and distinctly Cornish identity, both in their industrial practice and in their social relations. Today, all over the world in areas where Cornish miners set up shop, one finds Cornish-style mining infrastructure, Methodist churches, and pasties – a savory pastry filled with meat and vegetables (Payton 1999). Although they assimilated to their new regions in necessary ways – for example, replacing swede in their pasties with pumpkin in Australia – they were notable for retaining dialect and communal identity well into the twentieth century (Pryor 1969). As noted in Chapter 6, because of this strong cultural identity, as well as their marginal economic and geographical position in contemporary Europe, the Cornish are typically viewed as conservative and resistant to innovation.

Yet, in Australia – and particularly the South Australian Yorke Peninsula mining communities of Moonta, Kadina, and Wallaroo, which retain a strong Cornish identity to this day – communities of Cornish miners were highly creative, and indeed revolutionary (in its political meaning) (Payton 2007). The innovativeness of these Cornish miners contrasts strongly with the more innovation-resistant culture of Cornish mining communities elsewhere in the diaspora – and, indeed, in Cornwall. This innovativeness took many forms, and includes the local invention of new machinery, such as the automatic jig, invented and sold by Cornish-born, self-taught machinist Frederick May (chief engineer at the Moonta mines); new governance structures relating to the commodification of water in arid Australia (Lawrence and Davies 2015); and the emergence of new social and political orientations, particularly trade unionism (Davies 1995). Striking and unionism were not features of Cornish mining communities elsewhere in the world (Nauright 2005),[5] perhaps because mineral extraction was carried out according to the rather decentralized Cornish system of "tribute and tutwork," in which small groups of miners (frequently kin) bid for contracts to work sections of a mine, with a percentage of the ore raised returning to them as payment (Drew and Connell 1993). Nevertheless, in South Australia, Cornish miners were at the forefront of industrial action, fomenting general strikes at the mines in Burra in 1848, and Moonta in 1864 and 1874. Many with ties to Moonta and Burra played a foundational role in both the labor movement and the organized Labor Party in Australia (Payton 2016).

Yet, the source or cause of this innovativeness within the Cornish mining community in nineteenth-century South Australia has not been readily accounted for. An overview of the literature suggests we should look to familiar push–pull factors. Lawrence and Davies (2015) posit that Australia's intense aridity forced immigrant Cornish people to develop new technologies and governance strategies in order to maintain traditional mining practices. Indeed, while the environmental pressures faced by South Australian miners were high, financial pressures in some communities were quite a bit lower. After some early experiments with so-called Company Towns, Australian mining entrepreneurs abandoned the idea of providing housing to their employees (Bell 1998). Instead, miners often built their own cottages of freely available, scavenged material near established villages or on the mineral lease, as at Moonta and Wallaroo. Consequently, ad hoc villages sprang up around the heavy industry, comprising cottages with gardens, churches, and institutes (Franklyn 1881, 93–4) (Fig. 7.3). The result was that, unlike in other nineteenth-century mining communities, the miners on the Yorke Peninsula (and elsewhere, as at Burra, South Australia, where separate sub-divisions were surveyed and dugout homes were constructed in the river banks; Mullen and Birt 2009) were largely not beholden to landlords or mine bosses for their accommodation, a situation that one can imagine they found remarkably freeing, perhaps alleviating some of the social and financial risks associated with innovation and experimentation. Further, Philip Payton (2014, 2016) suggests that the emerging radical tradition had its roots in a distinctly Cornish version of Methodism, bolstered by a shared sense of ethnic identity within the diasporic community that lent itself to social solidarity. South Australia was (and remains) a far-off, distant place – geographically marginal not just to Cornwall or the Atlantic Cornish diaspora, but also to the eastern Australian colonies. The inland and Yorke Peninsula mining communities feel worlds away from Adelaide, the closest urban center and seaport. This marginality, as we saw in the previous section, may have contributed both to the ethnic solidarity of Cornish miners (Payton

Figure 7.3 Hamley Flat, Moonta Mines, *c.* 1895.

2001, 60–1) and to their perceived freedom to experiment, tinker, and enact new social and technological solutions for the problems they encountered and experienced.

Certainly, each of these elements likely played a role in fostering innovation and a sense of innovativeness in Australia's nineteenth-century mining communities. The environment and landscape were harsh, arid, and unmistakably foreign to Cornish-born miners (as well as to their Irish, German, English, Scottish, and Chinese colleagues). This would have provided the sort of "push" factors that economists and archaeologists look for in cases of technological change and innovation. "Pull" factors were also in play, since the unique social conditions of these geographically isolated mining communities, and particularly the community and ethnic cohesiveness that developed, seem to have under-pinned a sense of solidarity probably linked to the local interest in trade unionism. Nevertheless, singular, linear explanations for innovation and innovativeness are not just unsatisfying, they are also overly simple and cannot account for the complex lived experience of the people whose choices, experiments, strike actions, and political views made up the tapestry of this period. Instead we must look for a more complicated – or perhaps, better phrased, a more tangled – way of understanding innovativeness.

Demography, density, connectedness

One of the most influential archaeological models of innovativeness was articulated most clearly by evolutionary archaeologist Stephen Shennan (2001), who argued that cultural innovation processes – that is the diffusion of knowledge and techniques from one generation to the next and throughout a population – not only correlated with population size and density, but were positively impacted by them. In other words, the bigger and better connected a population, the more likely that useful innovations would be invented and spread, while the sparser a population, the more likely that these would be lost or forgotten. This model, like much of the work under the evolutionary archaeology umbrella, took inspiration from biological models of mutation and reproduction and brought these together with Boyd and Richerson's (1985) dual inheritance theory.

Shennan's model was rapidly adopted by selectionists interested in culture change and innovation, and particularly those concerned with the changing practices and populations of Pleistocene hominins (Lycett and Norton 2010; Powell *et al.* 2009). Over the subsequent decade, further and more elaborate models were produced that used demographic factors to explain the loss of technological knowledge (Henrich

2004), the importance of social connectedness for the dissemination of innovations (Stiner and Kuhn 2006), and technological stability or conservatism (Premo and Kuhn 2010). These high-profile publications and research groups argued, effectively, that larger populations had higher numbers of inventors *as well as* a more robust distribution network for new ideas and practices. Furthermore, these same apparent advantages also accrue to densely interconnected populations, meaning that smaller but well-connected populations could demonstrate equivalent innovativeness to larger groups (Fig. 7.4). Steven Kuhn (2012) applies this directly to the question of a possible Upper Palaeolithic revolution. He suggests that *Homo sapiens* society in this period might be best modeled as a "small world" network comprising small, tightly connected groups

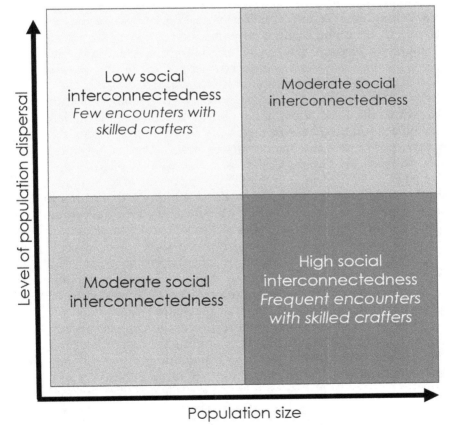

Figure 7.4   Model of interconnectedness based on population size and settlement density. Innovativeness in this model is a product of interconnection because this is assumed to pre-suppose frequency of contact with skilled crafters.

of twenty or thirty individuals with each member having weak ties to other groups: "a network structure that minimises average social distances [and] will maximise the rates at which 'contagions' such as novel cultural behaviours can spread" (Kuhn 2012, 77).

The demographic turn (*sensu* Collard *et al.* 2016) in evolutionary archaeology is not entirely a recent invention. Population pressure and demography are frequently invoked by historical behavioral ecologists as important causal factors in technological change. Richard Redding (1988), for example, brings together optimization models with archaeological and anthropological data to argue that hunter-gatherer groups engaged in cultivation in order to relieve the stress of a growing population on their wild resource base. In contrast to the more recent selectionist models, demographic changes, in this view, negatively impact fitness, leading to invention and adoption of innovations. Mellars (2005) has offered a sort of half-way position between ecological and selectionist conceptualizations of population pressure, suggesting that demographic changes drive increased interaction (and, hence, density of social connectedness) as well as competition for resources – a sort of combined push–pull factor for innovativeness.

Yet, this somewhat more complex version of the model reveals a key flaw within the assumptions underlying these demographic models: they assume a normative state of increasing social and technological complexity, positioning this uncritically as more adaptive, advantageous, and advanced. As Vaesen, Collard and colleagues (Collard *et al.* 2016; Vaesen *et al.* 2016) point out in their sharp critique of these models, the more complex a technology, the harder it is to use, explain, and repair, limiting its use and manufacture by a broad swath of the community. Their critique, moreover, addresses methodological aspects of the mathematical models applied by Joseph Henrich (2004) and Adam Powell *et al.* (2009), as well as an obvious reliance on erroneous and outdated archaeological and anthropological comparanda in constructing their models (cf. Read 2006).[6] Indeed, a recent test of these models by Alberto Acerbi *et al.* (2017) found that small population size was no obstacle to the retention, transmission, and replication of complex social information (*contra* Mellars 2006). They studied the diversity of folktales across a number of small and large population groups and compared several elements, including total number of tale types, variety of narrative motifs, and number variants of a given tale type. While they found that large populations had a larger number of folktales, these were, in fact, simpler in form and narrative motif than those recorded in smaller populations. They argue that larger populations may prefer more compressible stories that are accessible to and able to be learned by a greater number of individuals, and that increased numbers of folktales may simply represent the many regional traditions of a larger geographical area covered by a bigger population. They also point, tellingly, to the issue of sampling bias, warning that folktales from larger populations may be over-represented in their dataset because of differing research agendas and availability of data. This is a particularly relevant critique for archaeologists – especially those working on the remote past and early hominins. Since our dataset is small, fragmentary, and biased by taphonomic processes, a larger or denser population would be better represented, having produced more stuff (the inevitable result of more people making, using, and discarding things). Thus, the archaeological record will reflect more data (and more evidence of innovative behavior) from large populations than from smaller ones.

And yet, despite these obvious problems, I do not want to dismiss the demographic models entirely because I think they hold an incredibly valuable insight about innovativeness and how it emerges. Connectivity and population size are the core variables within these models, with palaeodemography calculated in a variety of complex ways (French 2016) and environmental factors being considered key constraints (Kuhn 2012; Richerson *et al.* 2009). This approach resonates with recent work in organizational and innovation studies that explore the possibilities of open or communal innovation. Eric von Hippel (2005, 94–6), for example, points to user innovations (more or less the application of the use innovativeness discussed above) as a domain where dispersed, heterogeneous communities of users, each with different experience and expertise as well as different needs, alter and adjust technologies in dialogue with each other. These "innovation communities" are loosely bound by a shared interest in using and adapting a given technology or system, but may include more social community elements, including support, a sense of belonging, or shared identity. The dynamism of these communities comes from the wide variety of backgrounds and needs large numbers of individuals and firms bring together. Building on this and other work around open innovation systems, Ferdinand and Meyer (2017, 4) propose that innovations necessarily emerge from ecosystems, "mutually intertwined social, economic, and material contexts" including individuals, communities, firms, their activities, and their relationships. Different knowledge sets are dispersed through different populations but are able to be brought together through their engagement in the wider innovation ecosystem. In other words, both innovation and innovativeness can be conceived of as emerging from and being sustained by heterogeneous networks of people, organizations, and knowledge sets. Indeed, considerable research has established that network diversity – in terms of both personal traits such as ethnicity, nationality, or gender, and also professional traits, such as skill-set or disciplinary speciality – leads to novelty, creativity, and innovation (Hofstra *et al.* 2020; Moser and San 2020).

As discussed in earlier chapters, neither the population in which people operate nor the network that connects them is necessarily purely human. Hodder (2012), for example, builds a model of human society in which the complex and long-term entanglement of people and things shapes reality and leads to social and technological changes. This work builds on a wide body of research across the social sciences, arguing that our social networks are heterogeneous in that they incorporate the material world, giving human–non-human relationships equal weight to human–human relationships (Knappett and Malafouris 2008; Latour 2005; Law 1987; Witmore 2007). Others have gone further, arguing that humans ourselves are not bounded entities but enmeshed inextricably with technology, ambulatory human–thing systems: cyborgs (Haraway 1985) or artificial apes (Taylor 2010). In recent years, a number of anthropologists have posited that the non-human elements in our network must be expanded to include the natural world we inhabit and with which we engage (Lestel *et al.* 2006; Todd 2014). Ingold (2013) makes the case that animals, as animate beings, are fundamental parts of our relational community, entwined with our emotions, intelligence, and experience of existence. Anna Tsing (2012), building on Haraway (2003), argues that plants and complex systems of plants – mushrooms, grains, sugar-cane plantations – have shaped humanity, human society, and our socio-political relations (with each other, with the material world, and with the environment). Even our bodies are entangled biological systems, in which our meat and bones engage constantly in symbiosis with myriad bacteria and small organisms whose DNA is entwined with ours (McFall-Ngai *et al.* 2013). In other words, the heterogeneous network in which we are entangled twines together not just people and things, but also the whole of the non-human world.

The link I want to make here is between the idea of connectedness and what Tsing (2012, 144) describes as "a human nature that shifted historically together with varied webs of interspecies dependence." We might take inspiration from the selectionists and argue that an increasingly connected world also connects people to place to thing to creature and to plant. Demographic pressure within that network emerges not just from an increased number of humans, but also from shifting proportions and types of these other-than-human elements. The non-humans within our networks, and especially those that make up the natural world, are often positioned as constraints to innovation, but I want to suggest they can also be perceived as provocations. New elements in the network – whether these be new hominins with whom to communicate (and procreate, our genetic history makes that clear), new foods to taste, new textures of sand or stone beneath our feet, new materials to shape into tools, new colors of stone or water, new weather to feel on our skin, new spaces to inhabit, new birds to listen to at night and marvel at during the day – should impact the whole of the network, as would, of course, the loss of old tools, beloved family, birthplaces, and the sounds of extinct animals. A network implies homoeostasis, but, being a mobile species, we exist in flux. Thus, our network should be constantly in the process of being re-shaped and newly constituted. We never stand still; why should it?

In a Pleistocene context, we can envision small groups of *H. sapiens* moving through new or simply foreign landscapes, tasting new fruit, and coming into contact (more or less regularly) with other familiar-but-different people. We know our hominin cousins overlapped and engaged with us because enough of them slept with enough of us that all *H. sapiens* are hybrids, bearing a small proportion of Neanderthal and Denisovan genetic material in their DNA. Perhaps it is no surprise that highly innovative regions such as southern and eastern Africa (birthplace of hominin toolmaking and symbolic behavior), Indonesia (home of the earliest-known rock art and origin of the first long-distance maritime voyagers), and southern Europe (where a variety of artistic and other symbolic behaviors flourished in both Neanderthal and *H. sapiens* populations) are also areas that potentially saw the longest overlap between our species and a hominin cousin or cousins. Our networks grew denser, we sought to understand their new form and our newly adjusted position within them, and reacted with creative exuberance. A counterpoint might be Australia, where creativity certainly was not lacking in the Pleistocene, but no other hominins were present, and the human population was sparse and widely distributed. However, the first humans to arrive in Australia encountered an alien world with radically different, deeply (probably unsettlingly) unfamiliar flora and fauna, which must have altered and enriched their networks and sense of self in complex and profound ways. That the art, cosmology, and culture of contemporary Aboriginal people is so deeply entwined with Country, place, and the natural world hints at the importance of these to the creative expression and innovative spirit of their ancestors. I would argue that we see a similar, though perhaps less intense, version of this occurring to and among the Cornish miners of nineteenth-century South Australia. The Yorke Peninsula was certainly much less densely settled by humans than was Cornwall, but these mining communities were marked by an incontrovertible burst of creative problem solving and innovative social practice. They too found themselves suddenly more densely connected, not just with each other through bonds of shared experience and perceived ethnicity, but also with more foreign and unknown elements than previously imagined, though only some of these were human.

So, in seeking to understand innovativeness, I argue we must look beyond the immediate apparent needs faced by a person or group. Those needs are socially conditioned and historically constituted: in hindsight, we can only guess at what past people felt were their most urgent problems. Fitness and efficiency, as I argued in Chapter 2, are not adequate for our purposes, being a product of late-twentieth-century social and economic thinking. Instead, we can frame innovativeness as emerging from complex, enmeshed networks of humans and other-than-humans. As these shift over time and through space – which they inevitably do – new relations are constituted, new practices inaugurated, and different elements conjoined into new forms. This state of flux is the conduit between a universally experienced creativity and a historically, geographically, and socially constituted innovativeness. Technological change, in particular, is often explained as humans applying force to materials to shape nature; but if both materials and nature are inextricable parts of humanity, we are going to have to shift our whole frame of reference. Instead, understanding innovativeness means looking holistically at the heterogeneous and complex networks from which innovations emerge, as well as their histories of growth and contraction.

Notes

1 Presumably, this is due, at least in part, to the advent of DNA analysis, which demonstrated quite clearly that we are all much more closely related than racist difference narratives would have it (Marks 2020).

2 Australian Museum, "*Homo sapiens*–modern humans," https://australianmuseum. net.au/learn/science/human-evolution/homo-sapiens-modern-humans/ (accessed September 18, 2020).

3 This literature is vast and deep, and far more extensive than I can possibly cite here. A good start to understanding creativity from the perspective of psychology and neuroscience – as well as the operationalization of this research in education, organizational studies, and beyond – can be found in recent handbooks on creativity research (Kaufman and Sternberg 2005, 2010; Rickards *et al.* 2009), as well as interdisciplinary journals such as the *Creativity Research Journal* and *Creativity and Innovation Management*.

4 "The killing times," www.theguardian.com/australia-news/ng-interactive/2019/ mar/04/massacre-map-australia-the-killing-times-frontier-wars (accessed September 18, 2020).

5 Although riots in response to food shortages occurred several times in Cornish history during the eighteenth and early nineteenth centuries (Payton 2014, 18).

6 Surprisingly, none of these critiques has addressed the assumption made by Henrich that his population of social learners/teachers comprises only male

adults, while the elderly, children, and women play no significant role in the transmission of knowledge or the spread and maintenance of innovative practices (Henrich 2004, 202). No wonder his population is maladaptive: it fails to function in any way like a genuine population of interconnected people – and, of course, is nothing at all like Aboriginal Australian communities, in which elders play highly significant roles in the conservation and dissemination of important cultural information.

Conclusion: The widening gyre

One hundred years ago, in the shadow of World War I and the Irish War of Independence, W. B. Yeats wrote his famous poem "The Second Coming." In this poem, he deploys a variety of natural, biblical, and even archaeological metaphors and images to emphasize the archetypal interwar period feeling that the world was coming undone; that a cusp had been reached; that the horrors just visited had unleashed something new, dynamic, and unquestionably threatening. Among the unformed threats that Yeats feared were slouching towards Bethlehem, we must include the weapons that made World War I such a bloodbath. After the unbridled, almost giddy technological optimism of the late nineteenth and early twentieth centuries, World War I represented an epically destructive technological bender. Electricity, cars, aeroplanes, new chemicals, and new scientific means of perceiving the world had been weaponized – literally in many cases – against the very people who had once celebrated their invention. The social structures and practice of conflict, and the technologies used to wage that conflict, were no longer in sync, and the result was death on a previously unimaginable scale. The falcon – new technologies and sciences, in this instance – certainly seemed to have flown out of earshot of its falconer.

It was during these same interwar years that economist Joseph Schumpeter began forming the theories of entrepreneurship, capitalism, and the business cycle that have become core texts in our innovation discourse. Erwin Dekker (2018) positions Schumpeter as a member of the *avant garde*: a Futurist fascinated by the dynamism and possibilities of radical technological change but rendered pessimistic and increasingly cautious by the calamity of World War I and subsequent socio-political developments. By the 1940s, Schumpeter had brought together this interest in dynamic models of development with observations drawn from Marxist scholarship and the contemporary political context to argue that innovation was both a source of economic growth and a "gale of creative destruction" that decimated pre-existing power structures, knowledge systems, and organizations. Of course, as discussed in Chapter 6, instead of understanding this observation as a critique of capitalism and a warning about its long-term viability, mainstream economists and business scholars have embraced it. Since the 1970s – and more so during the tech bubbles of the 1990s and 2000s – disruption has been championed as a method for social development and technological advancement in its own right.

In 2016, two years after I first outlined this book and well into its drafting stage, a report in the *Wall Street Journal* noted that "The Second Coming" was being cited in news reports and commentary pieces with unprecedented frequency (Ballard 2016). Much like Yeats' unease of a century before, it seems the tipsy drive to innovate and improve our technological infrastructure was giving way to an anxious hangover. The technologies at whose development we had marveled and for whose growth we had cheered in the previous decades – online stores such as Amazon and social media platforms such as Facebook – were becoming sinister. The tech-utopianism of the 1990s and early 2000s was succumbing to a more critical stance. Indeed, recent years have made clear that, *contra* the optimistic theories of prominent technology-studies scholars (e.g., Feenberg 1995, 2010), ground-breaking technologies seem most responsive to protest in bad faith (e.g., Twitter's broken reporting algorithms, which see women and non-white people lose their accounts for pushing back against white supremacists) and hacking in the form of monetization (i.e., the cynical green-washing and hunt for pink dollars that make corporations seem progressive in their response to climate change and LGBT inclusiveness but largely benefit only their bottom line). Moreover, we have seen complex and important research demonstrating that artificial intelligence systems incorporate the stereotypes and biases of the dominant society in which they are created, reinforcing the oppression of minority populations (Noble 2018). Similarly, genetic advances, once hailed as the last nails in the coffin for racial stereotypes, seem to feed into race thinking as much as they provide critiques of it (Wailoo *et al.* 2012). Mark Zuckerberg once exhorted us to "move fast and break things," but with the recent revelations regarding election hacking, data gathering, and racist radicalization facilitated by Zuckerberg's Facebook, among other tech corporations, it is fair to say we have started to ask what exactly it is we are breaking and who actually benefits. Suffice to say, in 2019 (and even more so in 2020), much as in 1919, there is a pervasive sense that the falcon – perhaps represented today by the icon of a small, blue bird – is again not able to hear the falconer.

Archaeology, though, seems still to be caught up in technological determinism and tech-utopianism. The hunt for Oldests and Firsts has not abated. Indeed, archaeologists often lag years or decades behind our sibling disciplines when it comes to theory. We did not discover feminism until the 1980s, nor actor-network theory until the 2000s. The complexities of our *bricolage* approach require us to take time to assimilate and piece together our bits of the past with bodies of theory; and we rely heavily on colleagues in adjacent fields to help us interpret and apply complicated social models developed with the contemporary world in mind. In drafting this manuscript, I have felt myself pulled back and forth between the contemporary world – where technology, innovation, globalization, and capitalism are all bound up in climatic catastrophes, increasingly oppressive and inescapable surveillance by multinational companies and political entities, and the return of fascism and white supremacy to the mainstream – and the past – where the confluence of power, dominance, and innovation is frequently less obvious. Indeed, I would suggest that our discipline-wide fixation on Oldests and Firsts is one means of avoiding complex model building and even more complex political conversations regarding how and why people engaged with novelty, developed creative practices, and adopted or rejected new practices. If I have made anything clear in the course of this book, I hope I have made it obvious how profoundly unproductive I find this approach.

There can be no argument that the past holds power. Archaeology and nationalism are closely wedded. Ethnic groups frequently anchor their claims to sovereignty in archaeological data – sometimes heavily manipulated. Indigenous peoples use archaeological discoveries – and sometimes collaborate directly with archaeologists – to push back against colonial narratives of conquest, collapse, and tenure. In the present moment, archaeologists and historians are grappling with how to respond when white supremacists and ethnonationalists claim specific past people, haplotypes, sites, and symbols as central to their identities and movements. As I have argued elsewhere, stories of technological change and innovation in the past also hold power in the present.[1] It strikes me as no accident that Yeats invokes the awakening of an ancient monument, the Great Sphinx of Giza, as a symptom of existential disruption and a sign of the end of the world. The past is alive in the present and can itself serve to disrupt and endanger the status quo. It is up to us to give shape to that disruption, to engage with the past in the present, and to tell stories about and build models of social change and technological development that acknowledge both the ideals of the future and the deep roots these innately human phenomena have in the past.

Indeed, the power of archaeological data is that they offer alternative worlds and ways of being to contrast with the contemporary world. The past is built from fragments in the present and, as such, can never be fully disentangled from contemporary assumptions, political structures, problems, and social modalities. However, its very fragmentation offers us endless possibilities to test ideas about human relationships, social structures, systems of trade, and patterns of communication. We do not need – and in fact should not desire – to attempt a teleological reconstruction of past worlds with the contemporary system of globalized capitalism as its inevitable culmination. This artificially narrows the scope of the stories we can tell and the data we might consider. However, saying this is easier than enacting it. A key challenge faced by archaeologists in interpreting the archaeological record lies in disentangling our socially constituted expectations from our interpretative framework. That is, we are forced to challenge our own expectations about how the world works and might have worked because these are inextricably colored by our own present-day experiences. To put this another way: when we talk about innovation in the past using terminology, theoretical models, and patterns of behavior drawn from sociology, economics, and anthropology, we run the risk of uncritically reproducing the social and economic modalities of the last 200 years in Olde Worlde drag.

The social world of innovation

In order to push back against what I see as a tendency towards overly presentist arguments in archaeological studies of innovation, I have chosen to focus throughout this book on social relationships and social phenomena. The innovation-studies literature – and indeed the wider popular discourse around innovation – is profoundly bound up in common wisdom about social and economic relations within capitalist systems. Innovation narratives emphasize hierarchy, linear progress, monetization, and alienated labor. Nowhere is this clearer than in the undying myth of the solitary genius inventor whose creative spark emerges, it seems, from the void. Instead, as we have seen repeatedly throughout this book, not only is creativity socially structured and framed by human relations, invention itself is a social process in which individual acts of creation emerge from histories of practice, collaboration, iteration, and experimentation that implicate a web of relations among people, things, ideas, and practices. That both the lone genius and the iconic invention are bound up in the discourse of colonialism and oppression – here in Australia, the reactionary right like to tout the invention of the wheel as some sort of trump card in favor of white supremacy – is reason enough to find other narratives to champion.

Work by feminist anthropologists has highlighted how the sterile and rational world of capital, debt, exchange, and production is itself a mask hiding a rather more scuffed and complex, profoundly human mess. Tsing (2015), for example, makes clear that, even in the context of twenty-first-century globalized supply systems moving valued goods (in this case, matsutake mushrooms) between source and vendor, the goods themselves are unstable, being both alienated commodities and entangled in social relations. Similarly, Caroline Schuster's (2015) study of microfinance in urban Paraguay traces the complex web of social and financial relationships that bind together women's microcredit groups and shape their relations with each other, with their communities, and with the lending NGOs. In the social economies they describe, power flows between people, things, and places, coalescing around sometimes surprising constellations of social relationships, practices, and technologies.

The value of centering these messy networks in a study of innovation is that it has allowed me to escape some of the least productive and most obfuscatory demons that haunt our narratives of change over time. Throughout the previous chapters, I have extensively critiqued evolutionary approaches to innovation, especially those co-embedded with optimization models. While I recognize the synthetic power of these approaches – in particular their ability to draw together archaeological data at a number of different scales and temporalities with ecological information – I find them fundamentally flawed in their drive to excise the messy complexity of human relationships in favor of more mechanistic and animalistic external processes. Furthermore, their predominant deployment by largely white anglophones to explain the past (sometimes quite recent) of colonized peoples – Native Americans, Aboriginal Australians, Pacific Islanders – echoes dehumanizing, colonial knowledge systems. These models center populations, not people, and erase the axes of power and relation through which, I have argued, innovations are communicated, negotiated, tested, and enacted. Moreover, they externalize and alienate innovations, denying their entanglement with persons, histories, relationships, and extended networks of technology and practice. Although we necessarily segment our research areas into manageable portions, we must acknowledge that innovation diffusion, for example, is inextricably imbricated and entangled with patterns of mobility, including the movement of marriage partners and their natal kin, patterns of teaching and learning, ritual and political systems that structure the transmission of knowledge, personal and communal histories of manufacture and practice that shape the adoption of new ideas and things, wider economic and social relationships between and among communities and kin groups, and environmental considerations, at the very least. Some of these variables will inevitably be unknown, others assuredly unknowable; but each is worth our consideration since they represent the residue of people's beliefs, actions, and engagements with each other and the wider world.

I started this book asking why innovation was worth studying and why archaeological investigations of innovation matter. In answer, I have presented a variety of case studies, examples, interpretations, and hypotheses drawing on data from our earliest hominin ancestors right up to the recent past. Each of these has allowed me to explore elements of the social world of innovation – from the invention process to broader considerations of tradition, conservatism, innovativeness, and creativity. These examples, like much of human practice, have at times been contradictory. What are we meant to make of a case study of conservative Romano-Cornish ceramicists when later in the same chapter – and more emphatically in the subsequent – tradition and innovation are proposed to be less oppositional states and more like a mobius strip: opposite and yet identical, each enfolded in the other?

In fact, there is no unifying theory in this book except its initial premise: people exist in complicated social constellations and, since innovation is a human practice, it too must be complicated by people and their interactions with each other; with the past; with the future; and with the tangible, material world. I see social models of innovation as a means to escape uncritically gendered and linear narratives of technological change. Archaeological data are ideal for this as, being fragmented, they offer more scope for exploration as we try to make meaning from seemingly disconnected sherds. This fragmentation offers us an almost infinite choice of perspectives from which to examine a given thing, practice, or social conformation, furnishing equally endless means of challenging the established discourse around innovation, technological change, and the social and economic relationships bound up in them. It offers us space to escape the common wisdom of the present; to try out new, radical conformations of person, place, and thing; and to test whether they change our vision of the wider world. As many of the case studies illustrated, a shift in perspective can fundamentally alter our interpretation of a whole period of history or suite of practices – whether this concerns the creativity of Aboriginal Tasmanians or emerging models of widespread and regular plant domestication.

Stories of innovation are often stories of norms and elites. This is a discourse of the dominant in which the peripheral, the laggards, the resistant are steamrolled by the inevitable changes coming their way. They and their non-conforming ways of life are impediments to be overcome. Yet a social-archaeological approach, one that centers the people whose history is not recorded but is only apparent in bits of broken pottery or discarded flint, must grapple with the friction, filters, and frontiers they introduce into the innovation process. As such, our descriptions of this process are richer, more complicated, and inevitably more messily human than a straightforward financial model or an evolutionary algorithm. Indeed, resistance is creative, and innovation is shaped by friction, just as a wooden plank can be radically re-formed with a plane and some sandpaper. These processes allow innovations to be fitted into pre-existing social worlds without catastrophically disrupting them, although inevitably their conformation also will shift to reflect new balances of power or forms of engagement. In the same way, imitation and iteration are also bound up in inventiveness rather than opposed to it, since they represent the historical trajectory of the innovation process. Innovation *can* be innately destructive of earlier patterns or organizations, but it need not be. Oppositional binaries may lead to compellingly grand narratives, but reality unspools at a much smaller scale and with many more people involved.

Vere Gordon Childe, in many ways the father of archaeological innovation studies, with his interest in revolutions and social change, might object that this sort of approach, built around assumptions of contingent development and recursive iteration, is overly focused on the trees and neglects the forest (cf. Robb and Pauketat 2013). What of the revolutions he spent his career delineating? However, I would argue back that revolutions, like innovations, are innately social, being embedded within and emergent from human and supra-human networks rather than driven by extrasomatic changes in technology or environment. Technology, of course, is also innately human. Even the most complex artificial intelligence no more exists outside human relationships and knowledge systems than a sandwich in an automat could prepare itself or appear in its vending machine without human intercession.

For a given potter living in the pre-colonial American southwest, her marriage into a community tens or hundreds of kilometers distant from her birth family might have felt as, or more, revolutionary than the addition of a few new motifs or gestures to her pottery-making repertoire. These revolutions, too, are worth exploring because they have also contributed in meaningful ways to the invention, adoption, and communication of innovations. If we are to move away from "male heroes, big projects and important organisations" (Wajcman 2000, 453) so that our studies of innovation can encompass a more holistic and complicated world – the one in which most people live – then we must perhaps re-imagine what it means for something to be revolutionary or innovative. This means not that we can only study the small-scale and domestic, but that our models should tack back and forth between scales – allowing the marriage of one potter into a distant community to inform our larger story of changing ceramic technologies.

So many assumptions have been shoved into the black box of Innovation that unpacking them can sometimes feel like unleashing a jack-in-the-box sprung on razor wire. In the previous chapters, I have attempted to build a model of innovation that does not just allow for, but is best understood through, the actions of women and children, Indigenous peoples, and occupied populations. My archaeology of innovation is a *bricolage* that juxtaposes those bits and pieces of their lives not erased or overwritten by the dominant discourse with tangible things, environmental models, broader patterns of change over time, and curious incidences of conservatism. I've reconstructed a vessel for you to examine – but I won't be offended if, at some time in the future, you take it to pieces and put it back together in a different shape.

Note

1 C. J. Frieman, "Innovation, continuity and the punctuated temporality of archaeological narratives," in A. McGrath, L. Rademaker, and J. Troy (eds.), *Everywhen: Knowing the Past through Language and Culture* (Lincoln: University of Nebraska Press, 2023).

References

Abidin, C. 2015. "Communicative ♥ intimacies: Influencers and perceived interconnectedness." *Ada: A Journal of Gender, New Media and Technology* 8. https://adanewmedia.org/2015/11/issue8-abidin/ (accessed September 16, 2020).

Abidin, C. 2016a. "'Aren't these just young, rich women doing vain things online?': Influencer selfies as subversive frivolity." *Social Media + Society* 2(2). DOI: 10.1177/2056305116641342.

Abidin, C. 2016b. "Visibility labour: Engaging with influencers' fashion brands and #OOTD advertorial campaigns on Instagram." *Media International Australia* 161(1): 86–100. DOI: 10.1177/1329878X16665177.

Acerbi, A., Kendal, J., and Tehrani, J. J. 2017. "Cultural complexity and demography: The case of folktales." *Evolution and Human Behavior* 38(4): 474–80. DOI: 10.1016/j.evolhumbehav.2017.03.005.

Adas, M. 1989. *Machines as the Measure of Men: Science, Technology and Ideologies of Western Dominance*. Ithaca, NY: Cornell University Press.

Alsos, G. A., Hytti, U., and Ljunggren, E. 2013. "Gender and innovation: State of the art and a research agenda." *International Journal of Gender and Entrepreneurship* 5(3): 236–56. DOI: 10.1108/IJGE-06-2013-0049.

Ames, K. 2007. "The archaeology of rank." In R. A. Bentley, H. D. G. Maschner, and C. Chippendale (eds.), *Handbook of Archaeological Theories*. Lanham, MD: AltaMira Press. 487–513.

Ames, K. M. 1991. "The archaeology of the longue durée: Temporal and spatial scale in the evolution of social complexity on the southern Northwest Coast." *Antiquity* 65(249): 935–45. DOI: 10.1017/S0003598X0008073X.

Ammerman, A. J., and Cavalli-Sforza, L. L. 1971. "Measuring the rate of spread of early farming in Europe." *Man* 6(4): 674–88. DOI: 10.2307/2799190.

Ammerman, A. J., and Cavalli-Sforza, L. L. 1973. "A population model for the diffusion of early farming in Europe." In C. Renfrew (ed.), *The Explanation of Culture Change*. London: Duckworth. 343–57.

Ammerman, A. J., and Cavalli-Sforza, L. L. 1984. *The Neolithic Transition and the Genetics of Populations in Europe*. Princeton: Princeton University Press.

Anthony, D. W. 1990. "Migration in archeology: The baby and the bathwater." *American Anthropologist* 92(4): 895–914.

Apel, J. 2001. *Daggers, Knowledge and Power*. Uppsala: Coast to Coast.

Appleton-Fox, N. 1992. "Excavations at a Romano-British round: Reawla, Gwinear, Cornwall." *Cornish Archaeology* 31: 69–123.

Archibugi, D., and Iammarino, S. 1999. "The policy implications of the global-isation of innovation." *Research Policy* 28(2–3): 317–36. DOI: 10.1016/S0048-7333(98)00116-4.

Arnold, J. A. 2012. "Detecting apprentices and innovators in the archaeo-logical record: The shell bead-making industry of the Channel Islands." *Journal of Archaeological Method and Theory* 19: 269–305. DOI: 10.1007/s10816-011-9108-1.

Arthur, K. W. 2018. *The Lives of Stone Tools: Crafting the Status, Skill, and Identity of Flintknappers.* Tucson: University of Arizona Press.

Arvela, P. 2013. "Ethnic food: The other in ourselves." In D. Sanderson and M. Crouch (eds.), *Food: Expressions and Impression.* Oxford: Inter-Disciplinary Press. 45–56.

Asheim, B. T., and Gertler, M. S. 2009. "The geography of innovation: Regional innovation systems." In J. Fagerberg and D. C. Mowery (eds.), *The Oxford Handbook of Innovation.* Oxford: Oxford University Press. 291–317.

Bailey, D. W. 2000. *Balkan Prehistory: Exclusion, Incorporation and Identity.* London: Routledge.

Bajic-Hajdukovic, I. 2013. "Food, family, and memory: Belgrade mothers and their migrant children." *Food and Foodways* 21(1): 46–65. DOI: 10.1080/07409710.2013.764787.

Ballard, E. 2016. "Terror, Brexit and US election have made 2016 the year of Yeats." *Wall Street Journal.* August 23, 2016. www.wsj.com/articles/terror-brexit-and-u-s-election-have-made-2016-the-year-of-yeats-1471970174 (accessed October 20, 2020).

Balme, J., Davidson, I., McDonald, J., Stern, N., and Veth, P. 2009. "Symbolic behaviour and the peopling of the southern arc route to Australia." *Quaternary International* 202(1): 59–68. DOI: 10.1016/j.quaint.2008.10.002.

Bamforth, D. B. 2002. "Evidence and metaphor in evolutionary archaeology." *American Antiquity* 67(3): 435–52. DOI: 10.2307/1593821.

Bandini, E., and Harrison, R. A. 2020. "Innovation in chimpanzees." *Biological Reviews.* DOI: 10.1111/brv.12604.

Bar-Yosef, O. 2002. "The Upper Paleolithic revolution." *Annual Review of Anthropology* 31(1):363–93.DOI:10.1146/annurev.anthro.31.040402.085416.

Bar-Yosef, O. 2011. "Climatic fluctuations and early farming in West and East Asia." *Current Anthropology* 52(S4): S175–S193. DOI: 10.1086/659784.

Bar-Yosef Mayer, D. E., and Porat, N. 2008. "Green stone beads at the dawn of agriculture." *Proceedings of the National Academy of Sciences* 105(25): 8548–51. DOI: 10.1073/pnas.0709931105.

Bar-Yosef Mayer, D. E., Porat, N., Gal, Z., Shalem, D., and Smithline, H. 2004. "Steatite beads at Peqi'in: Long distance trade and pyro-technology during the Chalcolithic of the Levant." *Journal of Archaeological Science* 31(4): 493–502. DOI: 10.1016/j.jas.2003.10.007.

Barker, D. 1995. "Economists, social reformers, and prophets: A feminist critique of economic efficiency." *Feminist Economics* 1(3): 26–39. DOI: 10.1080/714042247.

Barnett, H. G. 1953. *Innovation: The Basis of Cultural Change,* Problems of Civilisation. New York: McGraw-Hill.

Basalla, G. 1988. *The Evolution of Technology.* Cambridge: Cambridge University Press.

Bates, J., Petrie, C. A., and Singh, R. N. 2017. "Approaching rice domestication in South Asia: New evidence from Indus settlements in northern India." *Journal of Archaeological Science* 78: 193–201. DOI: 10.1016/j.jas.2016.04.018.

Bauer, M. 1995. "Resistance to new technology and its effects on nuclear power, information technology and biotechnology." In M. Bauer (ed.), *Resistance to New Technology: Nuclear Power, Information Technology and Biotechnology.* Cambridge: Cambridge University Press. 1–44.

Baxter, J. E. 2005. *The Archaeology of Childhood: Children, Gender, and Material Culture.* Walnut Creek, CA: AltaMira Press.

Bayman, J. 2009. "Technological change and the archaeology of emergent colonialism in the Kingdom of Hawai'i." *International Journal of Historical Archaeology* 13: 127–57. DOI: 10.1007/s10761-009-0076-z.

Beaudry, M. C., and Parno, T. G., eds. 2013. *Archaeologies of Mobility and Movement.* New York: Springer.

Beauvoir, S. de. 2009. *The Second Sex.* Trans. C. Borde and S. Malovany-Chevallier. London: Alfred A. Knopf.

Bedford, S., and Spriggs, M. 2008. "Northern Vanuatu as a Pacific crossroads: The archaeology of discovery, interaction, and the emergence of the 'ethnographic present.'" *Asian Perspectives* 47(1): 95–120. DOI: 10.1353/asi.2008.0003.

Bedford, S., Spriggs, M., Burley, D. V., Sand, C., Sheppard, P., and Summerhayes, G. 2019. "Debating Lapita: Distribution, chronology, society and subsistence." In S. Bedford and M. Spriggs (eds.), *Debating Lapita: Distribution, Chronology, Society and Subsistence.* Canberra: Australian National University Press. 5–36.

Bell, P. 1998. "Fabric and structure of Australian mining settlements." In A. B. Knapp, V. C. Piggott, and E. W. Herbert (eds.), *Social Approaches to an Industrial Past: The Archaeology and Anthropology of Mining.* London: Routledge. 27–38.

Berg, M. 2010. "The British product revolution of the eighteenth century." In J. Horn, L. N. Rosenband, and M. R. Smith (eds.), *Reconceptualizing the Industrial Revolution.* Cambridge, MA: MIT Press. 47–66.

Best, S. 2002. *Lapita: A View from the East,* Auckland: New Zealand Archaeological Association.

Bettinger, R. L., and Eerkens, J. 1999. "Point typologies, cultural transmission, and the spread of bow-and-arrow technology in the prehistoric Great Basin." *American Antiquity* 64(2): 231–42. DOI: 10.2307/2694276.

Bhabha, H. K. 1994. *The Location of Culture.* London: Routledge.

Bijker, W. E. 1987. "The social construction of Bakelite: Toward a theory of invention." In W. E. Bijker, T. P. Hughes, and T. J. Pinch (eds.), *The Social Construction of Technological Systems.* Cambridge, MA: MIT Press. 159–87.

Bijker, W. E. 1993. "Do not despair: There is life after constructivism." *Science, Technology and Human Values* 18(1): 113–38. DOI: 10.2307/689703.

Bijker, W. E. 1995. *Of Bicycles, Bakelites, and Bulbs: Toward a Theory of Sociotechnical Change.* Cambridge, MA: MIT Press.

Binford, L. R. 1968. "Post-Pleistocene adaptations." In L. R. Binford and S. R. Binford (eds.), *New Perspectives in Archaeology.* Chicago: Aldine. 313–41.

Binford, L. R. 1983. *In Pursuit of the Past: Decoding the Archaeological Record.* New York: Thames and Hudson.

Binford, L. R. 1992. "Seeing the present and interpreting the past and keeping things straight." In J. Rossignol and L. Wandsnider (eds.), *Space, Time, and Archaeological Landscapes.* New York: Plenum Press. 43–59.

Bintliff, J. L., ed. 1991. *The Annales School and Archaeology.* Leicester: Leicester University Press.

Bishop, J. C. 2014. " 'That's how the whole hand-clap thing passes on': Online/ offline transmission and multimodal variation in a children's clapping game." In C. Richards and A. Burn (eds.), *Children's Games in the New Media Age: Childlore, Media and the Playground.* London: Routledge. 53–84.

Blake, M. K., and Hanson, S. 2005. "Rethinking innovation: Context and gender." *Environment and Planning A* 37: 681–701. DOI: 10.1068/a3710.

Boden, M. A. 2004. *The Creative Mind: Myths and Mechanisms.* 2nd edn. London: Routledge.

Boone, J. L., and Smith, E. A. 1998. "Is it evolution yet? A critique of evolutionary archaeology." *Current Anthropology* 39(S1): S141–74. DOI: 10.1086/204693.

Bowdler, S., and Lourandos, H. 1982. "Both sides of Bass Strait." In S. Bowdler (ed.), *Coastal Archaeology in Eastern Australia: 1980 Valla Conference of Australian Archaeology.* Canberra: Department of Prehistory, Research School of Pacific and Asian Studies, Australian National University. 121–32.

Boyd, R., and Richerson, P. J. 1985. *Culture and the Evolutionary Process.* Chicago: University of Chicago Press.

Bradley, D., Kim, I., and Tian, X. 2017. "Do unions affect innovation?" *Management Science* 63(7): 2251–71. DOI: 10.1287/mnsc.2015.2414.

Bray, P. 2012. "Before ^{29}Cu became copper: Tracing the recognition and invention of metalleity in Britain and Ireland during the third millennium BC." In M. J. Allen, J. Gardiner and J. A. Sheridan (eds.), *Is There a British Chalcolithic? People, Place and Polity in the Later 3rd Millennium.* London: Prehistoric Society Research Papers. 418–34.

Bray, P., and Pollard, A. M. 2012. "A new interpretative approach to the chemistry of copper-alloy objects: Source, recycling and technology." *Antiquity* 86(333): 853–67. DOI: 10.1017/S0003598X00047967.

Brown, J. A. 1989. "The beginnings of pottery as an economic process." In S. E. van der Leeuw and R. Torrence (eds.), *What's New? A Closer Look at the Process of Innovation.* London: Unwin Hyman. 203–24.

Brown, J. S., and Duguid, J. 1991. "Organizational learning and communities-of-practice: Toward a unified view of working, learning, and innovation." *Organizational Science* 2(1): 40–57.

Brown, L. K., and Mussell, K. 1984. *Ethnic and Regional Foodways in the United States: The Performance of Group Identity.* Knoxville: University of Tennessee Press.

Bruland, K. 1995. "Patterns of resistance to new technologies in Scandinavia: An historical perspective." In M. Bauer (ed.), *Resistance to New Technology: Nuclear Power, Information Technology and Biotechnology.* Cambridge: Cambridge University Press. 125–46.

Bruland, K., and Mowery, D. C. 2006. "Innovation through time." In J. Fagerberg and D. C. Mowery (eds.), *The Oxford Handbook of Innovation.* Oxford: Oxford University Press. 349–79.

Brumfiel, E. M., and Earle, T. K., eds. 1987. *Specialization, Exchange, and Complex Societies.* New Directions in Archaeology. Cambridge: Cambridge University Press.

Bruner, E. M. 1993. "Epilogue: Creative persona and the problem of authenticity." In S. Lavie, K. Narayan, and R. Rosaldo (eds.), *Creativity/Anthropology.* Ithaca, NY: Cornell University Press. 321–34.

Buckley, W. F. 1955. "Our mission statement." www.nationalreview.com/art-icle/223549/our-mission-statement-william-f-buckley-jr (accessed September 17, 2020).

Bunn, S. 1999. "The nomad's apprentice: Different kinds of 'apprenticeship' among Kyrgyz nomads in Central Asia." In P. Ainley and H. Rainbird (eds.), *Apprenticeship: Towards a New Paradigm of Learning.* London: Kogan Page. 74–85.

Burmeister, S. 2000. "Archaeology and migration: Approaches to an archaeological proof of migration." *Current Anthropology* 41(4): 539–67. DOI: 10.1086/317383.

Cameron, D. 2003. "Gender issues in language change." *Annual Review of Applied Linguistics* 23: 187–201. DOI: 10.1017/S0267190503000266.

Carlson, W. B., and Gorman, M. E. 1990. "Understanding invention as a cognitive process: The case of Thomas Edison and early motion pictures, 1888–91." *Social Studies of Science* 20(3): 387–430. DOI: 10.1177/030631290020003001.

Carlyon, P. M. 1987. "Finds from the earthwork at Carvossa." *Cornish Archaeology* 26: 103–41.

Caro, T. M., and Hauser, M. D. 1992. "Is there teaching in nonhuman animals?" *Quarterly Review of Biology* 67(2): 151–74. DOI: 10.1086/417553.

Carotenuto, F., Tsikaridze, N., Rook, L., *et al.* 2016. "Venturing out safely: The biogeography of *Homo erectus* dispersal out of Africa." *Journal of Human Evolution* 95: 1–12. DOI: 10.1016/j.jhevol.2016.02.005.

Casterline, J. B., ed. 2001. *Diffusion Processes and Fertility Transition: Selected Perspectives.* Washington, DC.: National Academy Press.

Cath-Garling, S. 2017. *Evolutions or Revolutions? Interaction and Transformation at the "Transition" in Island Melanesia.* University of Otago Studies in Archaeology. Dunedin: University of Otago.

Cavalli-Sforza, L. L., and Feldman, M. W. 1981. *Cultural Transmission and Evolution: A Quantitative Approach.* Princeton: Princeton University Press.

Chapman, J. 2007. "The elaboration of an aesthetic of brilliance and colour in the Climax Copper Age." In F. Lang, C. Reinholdt, and J. Weilhartner (eds.), *Stephanos Aristeios: Archäologische Forschungen zwischen Nil und Istros. Festschrift für Stefan Hiller zum 65. Geburtstag.* Vienna: Phoibos. 65–74.

Chase-Dunn, C. K., and Hall, T. D., eds. 1991. *Core/Periphery Relations in Precapitalist Worlds.* Oxford: Westview.

Chatwin, B. 1987. *The Songlines.* London: Cape.

Chen, F.-C. 2013. "Cab cultures in Victorian London: Horse-drawn cabs, users and the city, *ca.* 1830–1914." Unpublished Ph.D. thesis, University of York.

Childe, V. G. 1925. *The Dawn of European Civilization.* London: A. A. Knopf.

Childe, V. G. 1929. *The Danube in Prehistory.* Oxford: Clarendon Press.

Childe, V. G. 1936. *Man Makes Himself.* London: Watts.

Chiu, S. 2012. "The way of doing things: What Lapita pottery can tell us about the stories of Austronesian expansion." *Journal of Austronesian Studies* 3(1): 1–25.

Chiu, S. 2019. "Measuring social distances with shared Lapita motifs: Current results and challenges." In S. Bedford and M. Spriggs (eds.), *Debating Lapita: Distribution, Chronology, Society and Subsistence.* Canberra: Australian National University Press. 307–34.

Choi, J. Y., Platts, A. E., Fuller, D. Q., Hsing, Y.-I., Wing, R. A., and Purugganan, M. D. 2017. "The rice paradox: Multiple origins but single domestication in Asian rice." *Molecular Biology and Evolution* 34(4): 969–79. DOI: 10.1093/molbev/msx049.

Christensen, C. M. 1997. *The Innovator's Dilemma: When New Technologies Cause Great Firms to Fail.* Boston, MA: Harvard Business School.

Christensen, C. M., and Eyring, H. J. 2011. *The Innovative University: Changing the DNA of Higher Education from the Inside Out.* San Francisco: Jossey-Bass.

Christensen, C. M., Horn, M. B., and Johnson, C. W. 2011. *Disrupting Class: How Disruptive Innovation Will Change the Way the World Learns.* 2nd edn. New York: McGraw-Hill.

Chu, N. 2018. "The paradoxes of creativity in Guangzhou, China's wholesale market for fast fashion." *Culture, Theory and Critique* 59(2): 178–92. DOI: 10.1080/14735784.2018.1443020.

Civáň, P., Ivaničová, Z., and Brown, T. A. 2013. "Reticulated origin of domesticated emmer wheat supports a dynamic model for the emergence of agriculture in the Fertile Crescent." *PLOS ONE* 8(11). DOI: 10.1371/journal.pone.0081955.

Clark, J. E., and Blake, M. 1994. "The power of prestige: Competitive generosity and the emergence of rank societies in lowland Mesoamerica." In E. M. Brumfiel and J. W. Fox (eds.), *Factional Competition and Political Development in the New World.* Cambridge: Cambridge University Press. 17–30.

Clarke, D. L. 1968. *Analytical Archaeology.* London: Methuen.

Clarkson, C., Jacobs, Z., Marwick, B., *et al.* 2017. "Human occupation of northern Australia by 65,000 years ago." *Nature* 547: 306. DOI: 10.1038/nature22968.

Clifford, J. 1989. "The others: Beyond the 'salvage' paradigm." *Third Text* 3(6): 73–8. DOI: 10.1080/09528828908576217.

Collard, M., Vaesen, K., Cosgrove, R., and Roebroeks, W. 2016. "The empirical case against the 'demographic turn' in Palaeolithic archaeology." *Philosophical Transactions of the Royal Society B: Biological Sciences* 371(1698). DOI: 10.1098/rstb.2015.0242.

Colley March, H. 1889. "The meaning of ornament." *Transactions of the Lancashire and Cheshire Antiquarian Society* 7: 161–92.

Cooke, P. 2001. "Regional innovation systems, clusters, and the knowledge economy." *Industrial and Corporate Change* 10(4): 945–74. DOI: 10.1093/icc/10.4.945.

Courel, B., Robson, H. K., Lucquin, A., *et al.* 2020. "Organic residue analysis shows sub-regional patterns in the use of pottery by northern European hunter-gatherers." *Royal Society Open Science* 7(4). DOI: 10.1098/rsos.192016.

Cox, A. 2005. "What are communities of practice? A comparative review of four seminal works." *Journal of Information Science* 31(6): 527–40. DOI: 10.1177/0165551505057016.

Coy, M. W., ed. 1989. *Apprenticeship: From Theory to Method and Back Again.* Albany: State University of New York Press.

Coyne, R. 1999. *Technoromanticism: Digital Narrative, Holism, and the Romance of the Real.* Cambridge, MA: MIT Press.

Craig, O. E., Steele, V. J., Fischer, A., *et al.* 2011. "Ancient lipids reveal continuity in culinary practices across the transition to agriculture in northern Europe." *Proceedings of the National Academy of Sciences* 108(44): 17910. DOI: 10.1073/pnas.1107202108.

Cripps, L. J. 2006. "In time and space: Sociotemporal relations, community and identity in Cornwall, 400 BC–AD 200." Unpublished Ph.D. thesis, University of Leicester.

Cronon, W. 2003. *Changes in the Land: Indians, Colonists, and the Ecology of New England.* Rev. edn. New York: Hill and Wang.

Crowden, N. 2003. *Examining Gender Bias in Studies of Innovation.* Centre for Policy Research on Science and Technology Reports. Vancouver: CPROST, Simon Fraser University. www.sfu.ca/cprost/?p=188.

Crown, P., and Wills, W. H. 1995. "The origins of southwestern ceramic containers: Women's time allocation and economic intensification." *Journal of Anthropological Research* 51(2): 173–86. DOI: 10.1086/jar.51.2.3630253.

Crown, P. L. 1994. *Ceramics and Ideology: Salado Polychrome Pottery.* Albuquerque: University of New Mexico Press.

Crown, P. L. 2001. "Learning to make pottery in the prehispanic American southwest." *Journal of Anthropological Research* 57(4): 451–69. DOI: 10.1086/jar.57.4.3631355.

Crown, P. L. 2002. "Learning and teaching in the prehispanic American southwest." In K. A. Kemp (ed.), *Children in the Prehistoric Puebloan Southwest.* Salt Lake City: University of Utah Press. 108–24.

Crown, P. L. 2016. "Secrecy, production rights, and practice within communities of potters in the prehispanic American southwest." In A. P. Roddick and A. B. Stahl (eds.), *Knowledge in Motion.* Tucson: University of Arizona Press. 67–96.

Cullen, B. 1993a. "The cultural virus." Unpublished Ph.D. thesis, University of Sydney.

Cullen, B. 1993b. "The Darwinian resurgence and the cultural virus critique." *Cambridge Archaeological Journal* 3(2): 179–202. DOI: 10.1017/S0959774300000834.

Cullen, B. 1995. "Living artefact, personal ecosystem, biocultural schizophrenia: A novel synthesis of processual and post-processual thinking." *Proceedings of the Prehistoric Society* 61: 371–91. DOI: 10.1017/S0079497X00003133.

Cullen, B. 1996a. "Cultural virus theory and the eusocial pottery assemblage." In H. D. G. Maschner (ed.), *Darwinian Archaeologies.* New York: Plenum Press. 43–65.

Cullen, B. 1996b. "Social interaction and viral phenomena." In J. Steele and S. Shennan (eds.), *The Archaeology of Human Ancestry: Power, Sex and Tradition.* London: Routledge. 378–89.

Danneels, E. 2004. "Disruptive technology reconsidered: A critique and research agenda." *Journal of Product Innovation Management* 21(4): 246–58. DOI: 10.1111/j.0737-6782.2004.00076.x.

David, B., McNiven, I. J., Richards, T., *et al.* 2011. "Lapita sites in the Central Province of mainland Papua New Guinea." *World Archaeology* 43(4): 576–93. DOI: 10.1080/00438243.2011.624720.

Davidson, I., and Roberts, D. A. 2008. "14,000 BP – on being alone: The isolation of Tasmania." In D. A. Roberts and M. Crotty (eds.), *Turning Points in Australian History.* Sydney: University of New South Wales Press. 18–31.

Davies, M. 1995. "Cornish miners and class relations in early colonial South Australia: The Burra Burra strikes of 1848–1849." *Australian Historical Studies* 26(105): 568–95.

Davis, F. D. 1989. "Perceived usefulness, perceived ease of use, and user acceptance of information technology." *MIS Quarterly* 13(3): 319–40. DOI: 10.2307/249008.

Davis, F. D., Bagozzi, R. P., and Warshaw, P. R. 1989. "User acceptance of computer technology: A comparison of two theoretical models." *Management Science* 35(8): 982–1003. DOI: 10.1287/mnsc.35.8.982.

Davis, F. D., Bagozzi, R. P., and Warshaw, P. R. 1992. "Extrinsic and intrinsic motivation to use computers in the workplace 1." *Journal of Applied Social Psychology* 22(14): 1111–32. DOI: 10.1111/j.1559-1816.1992.tb00945.x.

Davis, J. L. 2020. *How Artifacts Afford: The Power and Politics of Everyday Things*. Cambridge, MA: MIT Press.

Dawkins, R. 1976. *The Selfish Gene*. Oxford: Oxford University Press.

Dekker, E. 2018. "Schumpeter: Theorist of the avant-garde." *Review of Austrian Economics* 31(2): 177–94. DOI: 10.1007/s11138-017-0389-9.

Denham, T. 2009. "A practice-centered method for charting the emergence and transformation of agriculture." *Current Anthropology* 50(5): 661–7. DOI: 10.1086/605469.

Denham, T. 2011. "Early agriculture and plant domestication in New Guinea and Island Southeast Asia." *Current Anthropology* 52(S4): S379–95. DOI: 10.1086/658682.

Denham, T., Haberle, S. G., Lentfer, C., *et al.* 2003. "Origins of agriculture at Kuk Swamp in the highlands of New Guinea." *Science* 301(5630): 180–1. DOI: 10.1126/science.1085255.

d'Errico, F., and Stringer, C. B. 2011. "Evolution, revolution or saltation scenario for the emergence of modern cultures?" *Philosophical Transactions of the Royal Society B: Biological Sciences* 366(1567): 1060–9. DOI: 10.1098/rstb.2010.0340.

d'Errico, F., Doyon, L., Zhang, S., *et al.* 2018. "The origin and evolution of sewing technologies in Eurasia and North America." *Journal of Human Evolution* 125: 71–86. DOI: 10.1016/j.jhevol.2018.10.004.

Derricourt, R. M. 2018. *Unearthing Childhood: Young Lives in Prehistory*. Manchester: Manchester University Press.

Descola, P. 2014. "Modes of being and forms of predication." *HAU: Journal of Ethnographic Theory* 4(1): 271–80. DOI: 10.14318/hau4.1.012.

Dethlefsen, E., and Deetz, J. 1966. "Death's heads, cherubs, and willow trees: Experimental archaeology in colonial cemeteries." *American Antiquity* 31(4): 502–10. DOI: 10.2307/2694382.

Diamond, J. M. 1997a. *Guns, Germs, and Steel: The Fates of Human Societies*. New York: W. W. Norton.

Diamond, J. M. 1997b. "Location, location, location: The first farmers." *Science* 278 (5341): 1243–4. DOI: 10.1126/science.278.5341.1243.

Diamond, J. M. 2002. "Evolution, consequences and future of plant and animal domestication." *Nature* 418: 700–7. DOI: 10.1038/nature01019.

Diamond, J. M. 2005. *Collapse: How Societies Choose to Fail or Survive*. London: Allen Lane.

Díaz-Andreu García, M. 2007. *A World History of Nineteenth-Century Archaeology: Nationalism, Colonialism, and the Past*. Oxford: Oxford University Press.

Dickinson, W. R., and Shutler, R. 2000. "Implications of petrographic temper analysis for Oceanian prehistory." *Journal of World Prehistory* 14(3): 203–66. DOI: 10.1023/A:1026557609805.

Dobres, M.-A. 2000. *Technology and Social Agency: Outlining a Practice Theory for Archaeology.* Malden, MA: Blackwell.

Dogan, M. 1999. "Marginality." In M. A. Runco and S. R. Pritzer (eds.), *Encyclopedia of Creativity.* San Diego: Academic Press. 179–84.

Dornan, J. L. 2002. "Agency and archaeology: Past, present, and future directions." *Journal of Archaeological Method and Theory* 9(4): 303–29. DOI: 10.1023/A:1021318432161.

Dowson, T. A. 1994. "Reading art, writing history: Rock art and social change in southern Africa." *World Archaeology* 25(3): 332–45. DOI: 10.1080/00438243.1994.9980249.

Drew, G. J., and Connell, J. E. 1993. *Cornish Beam Engines in South Australian Mines*, special publication (South Australia, Department of Mines and Energy) no. 9. Adelaide: Department of Mines and Energy.

Dubb, S. 1999. *Logics of Resistance: Globalization and Telephone Unionism in Mexico and British Columbia*, Transnational Business and Corporate Culture. New York: Garland.

Dunbar, R. I. M. 2012. "Social cognition on the internet: Testing constraints on social network size." *Philosophical Transactions of the Royal Society B: Biological Sciences* 367(1599): 2192–201. DOI: 10.1098/rstb.2012.0121.

Dunnell, R. C. 1980. "Evolutionary theory and archaeology." *Advances in Archaeological Method and Theory* 3: 35–99. DOI: 10.2307/20170154.

Earle, T. 2004. "Culture matters in the Neolithic transition and emergence of hierarchy in Thy, Denmark: Distinguished lecture." *American Anthropology* 106: 111–25. DOI: 10.2307/3567446.

Earle, T., and Spriggs, M. 2015. "Political economy in prehistory: A Marxist approach to Pacific sequences." *Current Anthropology* 56(4): 515–44. DOI: 10.1086/682284.

Earle, T. K. 1987. "Chiefdoms in archaeological and ethnohistorical perspective." *Annual Review of Anthropology* 16: 279–308. DOI: 10.2307/2155873.

Earle, T. K. 1997. *How Chiefs Come to Power: The Political Economy in Prehistory.* Stanford: Stanford University Press.

Eckardt, H. 2014. *Objects and Identities: Roman Britain and the North-Western Provinces.* Oxford: Oxford University Press.

Eder, J. 2019. "Innovation in the periphery: A critical survey and research agenda." *International Regional Science Review* 42(2): 119–46. DOI: 10.1177/0160017618764279.

Edwards, L. 1997. "Myths over Miami." *Miami New Times.* June 5, 1997. https://web.archive.org/web/20200906020130/ (accessed September 16, 2020).

Ehrhardt, K. L. 2005. *European Metals in Native Hands: Rethinking the Dynamics of Technological Change, 1640–1683.* Tuscaloosa: University of Alabama Press.

Ehrhardt, K. L. 2009. "Copper working technologies, contexts of use, and social complexity in the eastern woodlands of native North America." *Journal of World Prehistory* 22: 213–35. DOI: 10.1007/978-1-4614-9017-3_13.

Ehrhardt, K. L. 2013. "'Style' in crafting hybrid material culture on the fringes of empire: An example from the Native North American midcontinent." In J. J. Card (ed.), *The Archaeology of Hybrid Material Culture.* Carbondale, IL: SIU Press. 364–96.

Elyachar, J. 2010. "Phatic labor, infrastructure, and the question of empowerment in Cairo." *American Ethnologist* 37(3): 452–64. DOI: 10.1111/j.1548-1425.2010.01265.x.

Ems, L. 2014. "Amish workarounds: Toward a dynamic, contextualized view of technology use." *Journal of Amish and Plain Anabaptist Studies* 2(1): 42–58. DOI: 10.18061/1811/59690.

Ems, L. 2015. "ICT non-use among the Amish." *Refusing, Limiting, Departing*. ACM CHI Conference on Human Factors in Computing Systems. Toronto, Canada. http://nonuse.jedbrubaker.com/wp-content/uploads/2014/03/CHI_EMS_NonUse_Workshopsmallpdf.com_.pdf (accessed September 17, 2020).

Epstein, S. R. 1998. "Craft guilds, apprenticeship, and technological change in preindustrial Europe." *Journal of Economic History* 58(3): 684–713. DOI: 10.2307/2566620.

Erez, M., and Nouri, R. 2010. "Creativity: The influence of cultural, social, and work contexts." *Management and Organization Review* 6(3): 351–70. DOI: 10.1111/j.1740–8784.2010.00191.x.

Erwin, D. H., and Krakauer, D. C. 2004. "Insights into innovation." *Science* 304 (5674): 1117–19. DOI: 10.1126/science.1099385.

Fagerberg, J. 2006. "Innovation: A guide to the literature." In J. Fagerberg, D. Mowery, and R. Nelson (eds.), *The Oxford Handbook of Innovation*. Oxford: Oxford University Press. 1–26.

Feenberg, A. 1992. "Subversive rationalization: Technology, power, and democracy." *Inquiry* 35(3–4): 301–22. DOI: 10.1080/00201749208602296.

Feenberg, A. 1995. "Subversive rationalization: Technology, power, and democracy." In A. Feenberg and A. Hannay (eds.), *Technology and the Politics of Knowledge*. Bloomington: Indiana University Press. 3–22.

Feenberg, A. 2002. *Transforming Technology: A Critical Theory Revisited*. 2nd edn. Oxford: Oxford University Press.

Feenberg, A. 2010. *Between Reason and Experience: Essays in Technology and Modernity*, Inside Technology. Cambridge, MA: MIT Press.

Fenn, T. 2015. "A review of cross-craft interactions between the development of glass production and the pyrotechnologies of metallurgy and other vitreous materials." *Cambridge Archaeological Journal* 25(1): 391–8. DOI: 10.1017/S0959774314001206.

Ferdinand, J.-P., and Meyer, U. 2017. "The social dynamics of heterogeneous innovation ecosystems: Effects of openness on community–firm relations." *International Journal of Engineering Business Management* 9: 1–16. DOI: 10.1177/1847979017721617.

Ferenstein, G. 2013. "Why labor unions and Silicon Valley aren't friends, in 2 charts." *TechCrunch*. https://techcrunch.com/2013/07/29/why-labor-unions-and-silicon-valley-arent-friends-in-2-charts/ (accessed September 17, 2020).

Fillitz, T., and Saris, A. J., eds. 2013. *Debating Authenticity: Concepts of Modernity in Anthropological Perspective*. New York: Berghahn.

Finlayson, C. 2019. *The Smart Neanderthal: Cave Art, Bird Catching, and the Cognitive Revolution*. Oxford: Oxford University Press.

Fischer, A. 1982. "Trade in Danubian shaft-hole axes and the introduction of a Neolithic economy in Denmark." *Journal of Danish Archaeology* 1: 7–12. DOI: 10.1080/010846X.1982.10589868.

Flath, C. M., Friesike, S., Wirth, M., and Thiesse, F. 2017. "Copy, transform, combine: Exploring the remix as a form of innovation." *Journal of Information Technology* 32(4): 306–25. DOI: 10.1057/s41265-017-0043-9.

Flexner, J. 2014. "Historical archaeology, contact, and colonialism in Oceania." *Journal of Archaeological Research* 22: 43–87. DOI: 10.1007/s10814-013-9067-z.

Fowler, C. 2004. *The Archaeology of Personhood: An Anthropological Approach.* London: Routledge.

Foxhall, L. 2000. "The running sands of time: Archaeology and the short-term." *World Archaeology* 31(3): 484–98. DOI: 10.2307/125114.

Franklyn, H. M. 1881. *A glance at Australia in 1880; or, Food from the south, showing the present condition and production of some of its leading industries, namely, wool, wine, grain, dressed meat.* Melbourne: Victorian Review.

Frei, K. M., Mannering, U., Kristiansen, K., *et al.* 2015. "Tracing the dynamic life story of a Bronze Age female." *Scientific Reports* 5. DOI: 10.1038/srep10431.

Frei, K. M., Mannering, U., Vanden Berghe, I., and Kristiansen, K. 2017a. "Bronze Age wool: Provenance and dye investigations of Danish textiles." *Antiquity* 91(357): 640–54. DOI: 10.15184/aqy.2017.64.

Frei, K. M., Villa, C., Jørkov, M. L., *et al.* 2017b. "A matter of months: High precision migration chronology of a Bronze Age female." *PLOS ONE* 12(6). DOI: 10.1371/journal.pone.0178834.

French, J. C. 2016. "Demography and the Palaeolithic archaeological record." *Journal of Archaeological Method and Theory* 23(1): 150–99. DOI: 10.1007/s10816-014-9237-4.

Frieman, C. J. 2010. "Imitation, identity and communication: The presence and problems of skeuomorphs in the Metal Ages." In B. V. Eriksen (ed.), *Lithic Technology in Metal Using Societies.* Aarhus: Jutland Archaeological Society. 33–44.

Frieman, C. J. 2012a. "Flint daggers, copper daggers and technological innovation in Late Neolithic Scandinavia." *European Journal of Archaeology* 15(3): 440–64. DOI: 10.1179/1461957112Y.0000000014.

Frieman, C. J. 2012b. "Going to pieces at the funeral: Completeness and complexity in British Early Bronze Age jet 'necklace' assemblages." *Journal of Social Archaeology* 12(3): 334–55. DOI: 10.1177/1469605311431400.

Frieman, C. J. 2012c. *Innovation and Imitation: Stone Skeuomorphs of Metal from 4th–2nd Millennia BC Northwest Europe.* Oxford: Archaeopress.

Frieman, C. J. 2013. "Innovation and identity: The language and reality of prehistoric imitation and technological change." In J. Card (ed.), *Hybrid Material Culture: The Archaeology of Syncretism and Ethnogenesis.* Carbondale, IL: Center for Archaeological Investigations. 318–41.

Frieman, C. J. 2020. "Revolutions so remote: Revolutionary thinking and archaeaological inquiry." *Age of Revolutions.* https://ageofrevolutions.com/2020/06/01/revolutions-so-remote-revolutionary-thinking-and-archaeological-inquiry/.

Frieman, C. J. 2023. "Innovation, continuity and the punctuated temporality of archaeological narratives." In A. McGrath, L. Rademaker, and J. Troy (eds.), *Everywhen: Australia and the Language of Deep History.* Lincoln, NE: University of Nebraska Press. 195–220.

Frieman, C. J., and Eriksen, B. V., eds. 2015. *Flint Daggers in Prehistoric Europe and Beyond.* Oxford: Oxbow.

Frieman, C. J., and Hofmann, D. 2019. "Present pasts in the archaeology of genetics, identity, and migration in Europe: A critical essay." *World Archaeology* 51(4): 528–45. DOI: 10.1080/00438243.2019.1627907.

Frieman, C. J., and Janz, L. 2018. "A very remote storage box indeed: The importance of doing archaeology with old museum collections." *Journal of Field Archaeology* 43(4): 257–68. DOI: 10.1080/00934690.2018.1458527.

Frieman, C. J., and Lewis, J. 2016. "Mountain barrows: A south-eastern Cornish barrow group in its local context." *Cornish Archaeology* 55: 145–61.

Frieman, C. J., and May, S. K. 2019. "Navigating contact: Tradition and innovation in Australian contact rock art." *International Journal of Historical Archaeology*. DOI: 10.1007/s10761-019-00511-0.

Fuller, D. Q. 2011. "Pathways to Asian civilizations: Tracing the origins and spread of rice and rice cultures." *Rice* 4(3): 78–92. DOI: 10.1007/s12284- 011-9078-7.

Fuller, D. Q. 2012. "New archaeobotanical information on plant domestication from macro-remains: Tracking the evolution of domestication syndrome traits." In P. Gepts, T. R. Famula, R. L. Bettinger, S. B. Brush, A. B. Damania, P. E. McGuire, and C. O. Qualset (eds.), *Biodiversity in Agriculture: Domestication, Evolution, and Sustainability*. Cambridge: Cambridge University Press. 110–35.

Fuller, D. Q., Willcox, G., and Allaby, R. G. 2011. "Cultivation and domestication had multiple origins: Arguments against the core area hypothesis for the origins of agriculture in the Near East." *World Archaeology* 43(4): 628–52. DOI: 10.1080/00438243.2011.624747.

Fuller, D. Q., Denham, T., Arroyo-Kalin, M., *et al.* 2014. "Convergent evolution and parallelism in plant domestication revealed by an expanding archaeological record." *Proceedings of the National Academy of Sciences* 111(17): 6147–52. DOI: 10.1073/pnas.1308937110.

Galeotti, A., and Goyal, S. 2009. "Influencing the influencers: A theory of strategic diffusion." *RAND Journal of Economics* 40(3): 509–32. DOI: 10.1111/j.1756-2171.2009.00075.x.

Gamble, C. 2012. "Creativity and complex society before the Upper Palaeolithic transition." In S. Elias (ed.), *Developments in Quaternary Sciences*. Amsterdam: Elsevier. 15–21.

Gamble, C., Gowlett, J., and Dunbar, R. 2011. "The social brain and the shape of the Palaeolithic." *Cambridge Archaeological Journal* 21(1): 115–36. DOI: 10.1017/S0959774311000072.

Gammage, B. 2013. *The Biggest Estate on Earth: How Aborigines Made Australia*. Sydney: Allen & Unwin.

Gaydarska, B., and Chapman, J. 2008. "The aesthetics of colour and brilliance – or why were prehistoric persons interested in rocks, minerals, clays and pigments?" In R. I. Kostov, B. Gaydarska, and M. Gurova (eds.), *Geoarchaeology and Archaeomineralogy: Proceedings of the International Conference, 29–30 October 2008 Sofia*. Sofia: St. Ivan Rilski. 63–6.

Gebauer, A. B., Sørensen, L. V., Taube, M., Kim, D. and Wieland, K.P. 2020. "First metallurgy in northern Europe: An Early Neolithic crucible and possible tuyère from Lønt, Denmark." *European Journal of Archaeology*. 24(1). DOI: 10.1017/ eaa.2019.73.

Gell, A. 1992. *The Anthropology of Time: Cultural Constructions of Temporal Maps and Images*. Oxford: Berg.

Gibson, C. 2010. "Guest editorial – creative geographies: Tales from the 'margins.' " *Australian Geographer* 41(1): 1–10. DOI: 10.1080/00049180903535527.

Gibson, C., Luckman, S., and Willoughby-Smith, J. 2010. "Creativity without Borders? Rethinking remoteness and proximity." *Australian Geographer* 41(1): 25–38. DOI: 10.1080/00049180903535543.

Gillin, P. 2007. *The New Influencers: A Marketer's Guide to the New Social Media.* Sanger, CA: Quill Driver.

Ginsburg, S., Jablonka, E., and Zeligowski, A. 2019. *The Evolution of the Sensitive Soul: Learning and the Origins of Consciousness.* Cambridge, MA: MIT Press.

Girard, R. 1990. "Innovation and repetition." *SubStance* 19 (2/62, 3/63), special issue, *Thought and Novation*, 7–20.

Godin, B. 2010a. " 'Innovation studies': The invention of a speciality (Part 1)." Project on the Intellectual History of Innovation. Working Paper 7. Montreal. www.csiic.ca.

Godin, B. 2010b. " 'Innovation studies': The invention of a speciality (Part 2)." Project on the Intellectual History of Innovation. Working Paper 8. Montreal. www.csiic.ca.

Godin, B. 2015a. *Innovation Contested: The Idea of Innovation over the Centuries.* London: Routledge.

Godin, B. 2015b. "Innovation: A conceptual history of an anonymous concept." Project on the Intellectual History of Innovation. Working Paper 21. Montreal. www.csiic.ca.

Godin, B., and Lucier, P. 2012. "Innovation and conceptual innovation in Ancient Greece." Project on the Intellectual History of Innovation. Working Paper 12. Montreal. www.csiic.ca.

Godin, B., and Vinck, D., eds. 2017. *Critical Studies of Innovation: Alternative Approaches to the Pro-Innovation Bias.* Cheltenham: Edward Elgar.

Golson, J. 1961. "Report on New Zealand, Western Polynesia, New Caledonia and Fiji." *Asian Perspectives* 5(1): 166–80.

Gooding, D. 1990. "Mapping experiment as a learning process: How the first electromagnetic motor was invented." *Science, Technology, & Human Values* 15(2): 165–201.

Gordon, R. J. 2017. *The Rise and Fall of American Growth: The US Standard of Living since the Civil War,* The Princeton Economic History of the Western World. Princeton: Princeton University Press.

Gorman, M. E., and Carlson, W. B. 1990. "Interpreting invention as a cognitive process: The case of Alexander Graham Bell, Thomas Edison, and the telephone. *Science, Technology, & Human Values* 15(2): 131–64. DOI: 10.1177/016224399001500201.

Gosden, C. 2004. *Archaeology and Colonialism: Cultural Contact from 5000 BC to the Present,* Topics in Contemporary Archaeology. Cambridge: Cambridge University Press.

Gosden, C. 2005. "What do objects want?" *Journal of Archaeological Method and Theory* 12(3): 193–211. DOI: 10.1007/s10816-005-6928-x.

Gosden, C., and Knowles, C. 2001. *Collecting Colonialism: Material Culture and Colonial Change.* Oxford: Berg.

Gott, B. 2002. "Fire-making in Tasmania: Absence of evidence is not evidence of absence." *Current Anthropology* 43(4): 650–6. DOI: 10.1086/342430.

Goulet, F., and Vinck, D. 2017. "Moving towards innovation through withdrawal: The neglect of destruction." In B. Godin and D. Vinck (eds.), *Critical Studies of Innovation: Alternative Approaches to the Pro-Innovation Bias.* Cheltenham: Edward Elgar. 97–114.

Gowlett, J. A. J. 2016. "The discovery of fire by humans: A long and convoluted process." *Philosophical Transactions of the Royal Society B: Biological Sciences* 371(1696). DOI: 10.1098/rstb.2015.0164.

Granovetter, M. S. 1973. "The strength of weak ties." *American Journal of Sociology* 78(6): 1360–80. DOI: 10.1086/225469.

Green, R. C. 1991. "Near and Remote Oceania: Disestablishing 'Melanesia' in culture history." In A. Pawley (ed.), *Man and a Half: Essays in Pacific Anthropology and Ethnobiology in Honour of Ralph Bulmer.* Auckland, NZ: Polynesian Society. 491–502.

Grider, S. A. 1980. "The study of children's folklore." *Western Folklore* 39(3): 159–69. DOI: 10.2307/1499798.

Grigoryan, L. K., Lebedeva, N., and Breugelmans, S. M. 2018. "A cross-cultural study of the mediating role of implicit theories of innovativeness in the relationship between values and attitudes toward innovation." *Journal of Cross-Cultural Psychology* 49(2): 336–52. DOI: 10.1177/0022022116656399.

Gron, K. J., and Sørensen, L. 2018. "Cultural and economic negotiation: A new perspective on the Neolithic transition of southern Scandinavia." *Antiquity* 92(364): 958–74. DOI: 10.15184/aqy.2018.71.

Guilford, J. P. 1950. "Creativity." *American Psychologist* 5(9): 444–54. DOI: 10.1037/h0063487.

Haak, W., Lazaridis, I., Patterson, N., *et al.* 2015. "Massive migration from the steppe was a source for Indo-European languages in Europe." *Nature* 522(7555): 207–11. DOI: 10.1038/nature14317.

Hägerstrand, T. 1966. "Aspects of the spatial structure of social communication and the diffusion of innovation." *Papers in Regional Science* 16: 27–42. DOI: 10.1007/BF01888934.

Hägerstrand, T. 1967. *Innovation Diffusion as a Spatial Process.* Chicago: University of Chicago Press.

Hall, B. H. 2006. "Innovation and diffusion." In J. Fagerberg and D. C. Mowery (eds.), *The Oxford Handbook of Innovation.* Oxford: Oxford University Press. 459–84.

Hall, E. R. 1946. "Zoological subspecies of man at the peace table." *Journal of Mammalogy* 27(4): 358–64. DOI: 10.2307/1375342.

Hall, T. D., and Chase-Dunn, C. 1993. "The world-systems perspective and archaeology: Forward into the past." *Journal of Archaeological Research* 1(2): 121–43. DOI: 10.1007/BF01326934.

Handler, R. 1986. "Authenticity." *Anthropology Today* 2(1): 2–4. DOI: 10.2307/3032899.

Handler, R., and Linnekin, J. C. 1984. "Tradition, genuine or spurious." *Journal of American Folklore* 97(385): 273–90. DOI: 10.2307/540610.

Harari, Y. N. 2014. *Sapiens: A Brief History of Humankind.* London: Harvill Secker.

Haraway, D.J. 1985. "A manifesto for cyborgs: Science, technology, and socialist feminism in the 1980s." *Socialist Review* 15(2): 65–107. DOI: 10.1080/08164649.1987.9961538.

Haraway, D. J. 2003. *The Companion Species Manifesto: Dogs, People, and Significant Otherness.* Chicago: Prickly Paradigm Press.

Härke, H. 1998. "Archaeologists and migrations: A problem of attitude? *Current Anthropology* 39(1): 19–46. DOI: 10.1086/204697.

Harris, O. J. T., and Cipolla, C. N. 2017. *Archaeological Theory in the New Millennium: Introducing Current Perspectives*. Abingdon: Routledge.

Harrison, R. 2002a. "Archaeology and the colonial encounter: Kimberley spearpoints, cultural identity and masculinity in the north of Australia." *Journal of Social Archaeology* 2(3): 352–77. DOI: 10.1177/1469605302002200304.

Harrison, R. 2002b. "Australia's iron age: Aboriginal post-contact metal arte- facts from Old Lamboo Station, southeast Kimberley, Western Australia." *Australasian Historical Archaeology* 20: 67–76. DOI: 10.2307/29544489.

Harrison, R. 2003. "The magical virtue of these sharp things." *Journal of Material Culture* 8(3): 311–36. DOI: 10.1177/13591835030083007.

Harrison, R. 2004. "Kimberley points and colonial preference: New insights into the chronology of pressure flaked point forms from the southeast Kimberley, Western Australia." *Archaeology in Oceania* 39(1): 1–11. DOI: 10.1102/ j.1834-4453.2004.tb0052.x.

Harrison, R. 2006. "An artefact of colonial desire? Kimberley points and the tech- nologies of enchantment. *Current Anthropology* 47(1): 63–88. DOI: 10.1086/ 497673.

Hawkes, C. 1954. "Archaeological theory and method: Some suggestions from the Old World." *American Anthropologist* 56(2): 155–68. DOI: 10.1525/ aa.1954.56.2.02a00660.

Hayden, B. 1995a. "A new overview of domestication." In T. D. Price and A. B. Gebauer (eds.), *Last Hunters, First Farmers: New Perspectives on the Transition to Agriculture*. Santa Fe: School of American Research. 273–300.

Hayden, B. 1995b. "Pathways to power: Principles for creating socioeconomic inequalities." In T. D. Price and G. M. Feinman (eds.), *Foundations for Social Inequality*. New York: Plenum. 15–78.

Hayden, B. 1998. "Practical and prestige technologies: The evolution of material systems." *Journal of Archaeological Method and Theory* 5(1): 1–55. DOI: 10.2307/20177377.

Hayden, B. 2001. "Richman, poorman, beggarman, chief: The dynamics of social inequality." In G. M. Feinman and T. D. Price (eds.), *Archaeology at the Millennium: A Sourcebook*. New York: Kluwer Academic. 213–68.

Hayden, B. 2014. *The Power of Feasts: From Prehistory to the Present*. Cambridge: Cambridge University Press.

Haydon, T., dir. 1978. *The Last Tasmanian*. Artis Film Productions.

Hayles, N. K. 1999. *How We Became Posthuman: Virtual Bodies in Cybernetics, Literature, and Informatics*. Chicago: University of Chicago Press.

Helms, M. W. 1988. *Ulysses' Sail: An Ethnographic Odyssey of Power, Knowledge, and Geographical Distance*. Princeton: Princeton University Press.

Helms, M. W. 1993. *Craft and the Kingly Ideal: Art, Trade, and Power*. Austin: University of Texas Press.

Henderson, J. C. 2007. *The Atlantic Iron Age: Settlement and Identity in the First Millennium BC*. London: Routledge.

Henrich, J. 2004. "Demography and cultural evolution: How adaptive cultural processes can produce maladaptive losses – The Tasmanian case." *American Antiquity* 69(2): 197–214. DOI: 10.2307/4128416.

Herring, P. 2009. "Early medieval transhumance in Cornwall, Great Britain." In J. Klapste (ed.), *Medieval Rural Settlement in Marginal Landscapes*. Turnhout, Belgium: Brepols. 47–56.

Hicks, D. 2010. "The material-cultural turn: Event and effect." In D. Hicks and M. C. Beaudry (eds.), *The Oxford Handbook of Material Culture Studies*. Oxford: Oxford University Press. 25–98.

Hicks, M. 2017. *Programmed Inequality: How Britain Discarded Women Technologists and Lost Its Edge in Computing*, History of Computing. Cambridge, MA: MIT Press.

Hilbert, L., Neves, E. G., Pugliese, F., *et al.* 2017. "Evidence for mid-Holocene rice domestication in the Americas." *Nature Ecology & Evolution* 1(11): 1693–8. DOI: 10.1038/s41559-017-0322-4.

Hindle, B. 1983. *Emulation and Invention*. New York: Norton.

Hirsch, B. T., and Link, A. N. 1987. "Labor union effects on innovative activity." *Journal of Labor Research* 8(4): 323–32. DOI: 10.1007/bf02685217.

Hirschman, E. C. 1980. "Innovativeness, novelty seeking, and consumer creativity." *Journal of Consumer Research* 7(3): 283–95. DOI: 10.2307/2489013.

Hiscock, P. 2007. *The Archaeology of Ancient Australia*. London: Routledge.

Hodder, I. 1982. *Symbols in Action: Ethnoarchaeological Studies of Material Culture*. Cambridge: Cambridge University Press.

Hodder, I. 1986. *Reading the Past: Current Approaches to Interpretation in Archaeology*. Cambridge: Cambridge University Press.

Hodder, I. 2012. *Entangled: An Archaeology of the Relationships between Humans and Things*. Oxford: Wiley-Blackwell.

Hofmann, D. 2016. "Keep on walking: The role of migration in Linearbandkeramik life." *Documenta praehistorica* 43: 235–51. DOI: 10.4312/dp.43.11.

Hoffmann, D. L., Angelucci, D. E., Villaverde, V., Zapata, J., and Zilhão, J. 2018a. "Symbolic use of marine shells and mineral pigments by Iberian Neandertals 115,000 years ago." *Science Advances* 4(2). DOI: 10.1126/sciadv.aar5255.

Hoffmann, D. L., Standish, C. D., García-Diez, M., *et al.* 2018b. "U-Th dating of carbonate crusts reveals Neandertal origin of Iberian cave art." *Science* 359(6378): 912–15. DOI: 10.1126/science.aap7778.

Hofstra, B., Kulkarni, V. V., Munoz-Najar Galvez, S., He, B., Jurafsky, D., and McFarland, D. A. 2020. "The diversity–innovation paradox in science." *Proceedings of the National Academy of Sciences* 117(17): 9284–91. DOI: 10.1073/pnas.1915378117.

Högberg, A. 2008. "Playing with flint: Tracing a child's imitation of adult work in a lithic assemblage." *Journal of Archaeological Method and Theory* 15(1): 112–31. DOI: 10.1007/s10816-007-9050-4.

Höppner, B., Bartelheim, M., Huijsmans, M., *et al.* 2005. "Prehistoric copper production in the Inn Valley (Austria), and the earliest copper in Central Europe." *Archaeometry* 47(2): 293–315. DOI: 10.1111/j.1475-4754.2005.00203.x.

Horning, A. 2015. "Comparative colonialism: Scales of analysis and contemporary resonances." In C. N. Cipolla and K. H. Hayes (eds.), *Rethinking Colonialism: Comparative Archaeological Approaches*. Gainesville: University of Florida Press. 234–46.

Hovers, E. 2012. "Invention, reinvention and innovation: The makings of Oldowan lithic technology." In S. Elias (ed.), *Developments in Quaternary Sciences*. Amsterdam: Elsevier. 51–68.

Hughes, T. P. 1987. "The evolution of large technological systems." In W. E. Bijker, T. P. Hughes, and T. J. Pinch (eds.), *The Social Construction of Technological Systems: New Directions in the Sociology and History of Technology.* Cambridge, MA: MIT Press. 51–82.

Hughes, T. P. 1994. "Technological momentum." In L. Marx and M. R. Smith (eds.), *Does Technology Drive History? The Dilemma of Technological Determinism.* Cambridge, MA: MIT Press. 101–13.

Ihuel, E., Pelegrin, J., Mallet, N., and Verjux, C. 2015. "The Pressigny phenomenon." In C. J. Frieman and B. V. Eriksen (eds.), *Flint Daggers in Prehistoric Europe and Beyond.* Oxford: Oxbow. 57–75.

Ingold, T. 2000. *The Perception of the Environment: Essays on Livelihood, Dwelling and Skill.* London: Routledge.

Ingold, T. 2013. "Anthropology beyond humanity." *Suomen antropologi: Journal of the Finnish Anthropological Society* 38(3): 5–23.

Irving, T. 2020. *The Fatal Lure of Politics: The Life and Thought of Vere Gordon Childe.* Melbourne: Monash University Publishing.

Isaksen, A., and Trippl, M. 2017. "Innovation in space: The mosaic of regional innovation patterns." *Oxford Review of Economic Policy* 33(1): 122–40. DOI: 10.1093/oxrep/grw035.

Jablonka, E., and Lamb, M. J. 1995. *Epigenetic Inheritance and Evolution: The Lamarckian Dimension.* Oxford: Oxford University Press.

Jablonka, E., Lamb, M. J., and Zeligowski, A. 2005. *Evolution in Four Dimensions: Genetic, Epigenetic, Behavioral, and Symbolic Variation in the History of Life,* Life and Mind: Philosophical Issues in Biology and Psychology. Cambridge, MA: MIT Press.

Johansen, K. L. 2006. "Settlement and land use at the Mesolithic–Neolithic transition in southern Scandinavia." *Journal of Danish Archaeology* 11: 201–23. DOI: 10.1080/0108464X.2006.10590118.

Johnson, M. 2010. *Archaeological Theory: An Introduction.* 2nd edn. Chichester: Wiley-Blackwell.

Johnson, V. 2007. "What is organizational imprinting? Cultural entrepreneurship in the founding of the Paris Opera." *American Journal of Sociology* 113(1): 97–127. DOI: 10.1086/517899.

Jones, A. L. 1993. "Exploding canons: The anthropology of museums." *Annual Review of Anthropology* 22: 201–20. DOI: 10.1146/annurev.an.22.100193.001221.

Jones, M. 1991. "Food production and consumption – plants." In R. T. J. Jones (ed.), *Roman Britain: Recent Trends.* Sheffield: Equinox. 21–8.

Jones, M., and Brown, T. 2000. "Agricultural origins: The evidence of modern and ancient DNA." *The Holocene* 10(6): 769–76. DOI: 10.1191/095968300095024.

Jones, R. 1971. "Rocky Cape and the problem of the Tasmanians." Unpublished Ph.D. thesis, Australian National University.

Jones, R. 1977. "The Tasmanian paradox." In R. V. S. Wright (ed.), *Stone Tools as Cultural Markers.* Canberra: Australian Institute of Aboriginal Studies. 189–204.

Jones, R. 1978. "Why did the Tasmanians stop eating fish?" In A. Gould (ed.), *Explorations in Ethnoarchaeology.* Albuquerque: University of New Mexico Press. 11–47.

Joordens, J. C. A., d'Errico, F., Wesselingh, F. P., *et al.* 2014. "*Homo erectus* at Trinil on Java used shells for tool production and engraving." *Nature* 518: 228. DOI: 10.1038/nature13962.

Jordan, P. 2015. *Technology as Human Social Tradition: Cultural Transmission among Hunter-Gatherers.* Oakland: University of California Press.

Jørgensen, L. B., Sofaer, J. R., and Sørensen, M. L. S. 2018. *Creativity in the Bronze Age: Understanding Innovation in Clay, Textile, and Metalwork Production.* Cambridge: Cambridge University Press.

Jude, G. 2014. "Return of the valley girl or unsung cyborg? (Beyond) media representations of glottal fry, a contentious US speaking practice." *Journal of Engaged Pedagogy* 13(1): 11–19.

Kardulias, P. N., ed. 1999. *World-Systems Theory in Practice: Leadership, Production, and Exchange.* Oxford: Rowman & Littlefield.

Kardulias, P. N., and Hall, T. D. 2008. "Archaeology and world-systems analysis." *World Archaeology* 40(4): 572–83. DOI: 10.1080/00438240802453252.

Kaufman, J. C., and Sternberg, R. J., eds. 2005. *The International Handbook of Creativity.* Cambridge: Cambridge University Press.

Kaufman, J. C., and Sternberg, R. J. 2010. *The Cambridge Handbook of Creativity*, Cambridge Handbooks in Psychology. New York: Cambridge University Press.

Keesing, R. M. 1982. "Kastom in Melanesia: An overview." *Mankind* 13(4): 297–301. DOI: 10.1111/j.1835-9310.1982.tb00994.x.

Keesing, R. M. 1984. "Traditionalist enclaves in Melanesia." In R. May and H. Nelson (eds.), *Melanesia: Beyond Diversity.* Canberra: Australian National University Press. 39–54.

Khamis, S., Ang, L., and Welling, R. 2017. "Self-branding, 'micro-celebrity' and the rise of social media influencers." *Celebrity Studies* 8(2): 191–208. DOI: 10.1080/19392397.2016.1218292.

Kienlin, T. L. 2008. "Tradition and innovation in Copper Age metallurgy: Results of a metallographic examination of flat axes from eastern Central Europe and the Carpathian Basin." *Proceedings of the Prehistoric Society* 74: 79–107.

Kienlin, T. L. 2010. *Traditions and Transformations: Approaches to Eneolithic (Copper Age) and Bronze Age Metalworking and Society in Eastern Central Europe and the Carpathian Basin*, BAR International Series. Oxford: Archaeopress.

Kienlin, T. L. 2014. "Aspects of metalworking and society from the Black Sea to the Baltic Sea from the fifth to the second millennium BC." In B. W. Roberts and C. P. Thornton (eds.), *Archaeometallurgy in Global Perspective: Methods and Syntheses.* New York: Springer. 447–72.

Kienlin, T. L. 2018. "On Europe, the Mediterranean and the myth of passive peripheries." In X.-L. Armada, M. Murillo-Barroso, and M. Charlton (eds.), *Metals, Minds and Mobility: Integrating Scientific Data with Archaeological Theory.* Oxford: Oxbow. 19–36.

Kienlin, T. L., and Ottaway, B. 1998. "Flanged axes of the North-Alpine region: An assessment of the possibilities of use–wear analysis on metal artefacts." In C. Mordant, M. Pernot, and V. Rychner (eds.), *L'atélier du bronzier en Europe du XXe au VIIIe siècle avant notre ère: Actes du colloque international "Bronze '96" Neuchâtel et Dijon, 1996.* Paris: CTHS. 271–86.

Killick, D., and Fenn, T. 2012. "Archaeometallurgy: The study of preindustrial mining and metallurgy." *Annual Review of Anthropology* 41(1): 559–75. DOI: 10.1146/annurev-anthro-092611-145719.

Killick, D. A. 2004. "Social constructionist approaches to the study of technology." *World Archaeology* 36(4): 571–8. 10.1080/0043824042000303746.

Kirch, P. V. 2017. *On the Road of the Winds: An Archaeological History of the Pacific Islands before European Contact.* 2nd edn. Berkeley: University of California Press.

Kirton, M. 1976. "Adaptors and innovators: A description and measure." *Journal of Applied Psychology* 61(5): 622–9. DOI: 10.1037/0021-9010.61.5.622.

Klassen, L. 2000. *Frühes Kupfer im Norden: Untersuchungen zu Chronologie, Herkunft und Bedeutung der Kupferfunde der Nordgruppe der Trichterbecherkultur,* Jutland Archaeological Society Publications 36. Højbjerg: Moesgård Museum and Jutland Archaeological Society; Aarhus University Press.

Klassen, L. 2002. "The Ertebølle culture and Neolithic continental Europe: Traces of contact and interaction." In A. Fischer and K. Kristiansen (eds.), *The Neolithisation of Denmark: 150 Years of Debate.* Sheffield: J. R. Collis. 305–17.

Klassen, L. 2004. *Jade und Kupfer: Untersuchungen zum Neolithisierungsprozess im westlichen Ostseeraum unter besonderer Berücksichtigung der Kulturentwicklung Europas 5500–3500 BC.* Aarhus: Jutland Archaeological Society; Moesgaard Museum.

Klassen, L., Pétrequin, P., and Grut, H. 2007. "Haches plates en cuivre dans le Jura français: Transferts à longue distance de biens socialement valorisés pendant les IVe et IIIe millénaires." *Bulletin de la Société préhistorique française* 104(1): 101–24. DOI: 10.2307/41549740.

Klein, R. G. 2000. "Archeology and the evolution of human behavior." *Evolutionary Anthropology: Issues, News, and Reviews* 9: 17–36. DOI: 10.1002/(SICI)1520-6505(2000)9:1<17::AID-EVAN3>3.0.CO;2-A.

Kline, S., and Rosenberg, N. 1986. "An overview of innovation." In R. Landau and N. Rosenberg (eds.), *The Positive Sum Strategy: Harnessing Technology for Economic Growth.* Washington, DC: National Academy Press. 275–304.

Knappett, C., and Malafouris, L., eds. 2008. *Material Agency: Towards a Non-Anthropocentric Approach.* New York: Springer.

Knight, M., Ballantyne, R., Robinson Zeki, I., and Gibson, D. 2019. "The Must Farm pile-dwelling settlement. *Antiquity* 93(369): 645–63. DOI: 10.15184/aqy.2019.38.

Kohl, P. L. 1987. "The use and abuse of world systems theory: The case of the pristine West Asian state." *Advances in Archaeological Method and Theory* 11: 1–35. DOI: 10.1016/B978-0-12-003111-5.50004-X.

Kohler, H.-P. 1997. "Learning in social networks and contraceptive choice." *Demography* 34(3): 369–83. DOI: 10.2307/3038290.

Krause, R. 2003. *Studien zur kupfer- und frühbronzezeitlichen Metallurgie zwischen Karpatenbecken und Ostsee.* Leidorf: Rahden/Westf.

Kristiansen, K. 1987. "From stone to bronze – the evolution of social complexity in northern Europe, 2300–1200 BC." In E. M. Brumfiel and T. K. Earle (eds.), *Specialization, Exchange, and Complex Societies.* Cambridge: Cambridge University Press. 30–51.

Kristiansen, K., and Larsson, T. B. 2005. *The Rise of Bronze Age Society: Travels, Transmissions and Transformations.* Cambridge: Cambridge University Press.

Kristiansen, K., Larsen, M. T., and Rowlands, M. J., eds. 1987. *Centre and Periphery in the Ancient World*, New Directions in Archaeology. Cambridge: Cambridge University Press.

Kristiansen, K., Allentoft, M. E., Frei, K. M., *et al.* 2017. "Re-theorising mobility and the formation of culture and language among the Corded Ware Culture in Europe." *Antiquity* 91(356): 334–47. DOI: 10.15184/aqy.2017.17.

Kroeber, A. L. 1940. "Stimulus diffusion." *American Anthropologist* 42(1): 1–20.

Kuhn, S. L. 2004. "Evolutionary perspectives on technology and technological change." *World Archaeology* 36(4): 561–70. DOI: 10.2307/4128289.

Kuhn, S. L. 2012. "Emergent patterns of creativity and innovation in early technologies." In S. Elias (ed.), *Developments in Quaternary Sciences*. Amsterdam: Elsevier. 69–87.

Kuhn, S. L., and Stiner, M. C. 1998. "Middle Palaeolithic 'creativity': Reflections on an oxymoron?" In S. J. Mithen (ed.), *Creativity in Human Evolution and Prehistory*. London: Routledge. 143–64.

Kuijpers, M. H. G. 2018. "The Bronze Age: A world of specialists? Metalworking from the perspective of skill and material specialization." *European Journal of Archaeology* 21(4): 550–71. DOI: 10.1017/eaa.2017.59.

Labov, W. 1990. "The intersection of sex and social class in the course of linguistic change." *Language Variation and Change* 2: 205–54. 10.1017/S0954394500000338.

Lake, M. 1998. "'Homo': The creative genus?" In S. J. Mithen (ed.), *Creativity in Human Evolution and Prehistory*. London: Routledge. 125–42.

Laland, K. N., and Reader, S. M. 2010. "Comparative perspectives on human innovation." In M. J. O'Brien and S. Shennan (eds.), *Innovation in Cultural Systems: Contributions from Evolutionary Anthropology*. Cambridge, MA: MIT Press. 37–52.

Lape, P. V. 2003. "A highway and a crossroads: Island Southeast Asia and culture contact archaeology." *Archaeology in Oceania* 38(2): 102–9. DOI: 10.1002/j.1834-4453.2003.tb00533.x.

Larsen, C. S. 1995. "Biological changes in human populations with agriculture." *Annual Review of Anthropology* 24: 185–213. DOI: 10.1146/annurev.an.24.100195.001153.

Latour, B. 1992. "Where are the missing masses? sociology of a few mundane artifacts." In W. Bijker and J. Law (eds.), *Shaping Technology/Building Society: Studies in Sociotechnical Change*. Cambridge, MA: MIT Press. 225–59.

Latour, B. 1993. *We Have Never Been Modern*. Cambridge, MA: Harvard University Press.

Latour, B. 2005. *Re-Assembling the Social: An Introduction to Actor-Network-Theory*. Oxford: Oxford University Press.

Lave, J. 1996. "Teaching, as learning, in practice." *Mind, Culture, and Activity* 3(3): 149–64. DOI: 10.1207/s15327884mca0303_2.

Lave, J. 2009. "The practice of learning." In K. Illeris (ed.), *Contemporary Theories of Learning: Learning Theorists … in Their Own Words*. London: Routledge. 229–37.

Lave, J. 2011. *Apprenticeship in Critical Ethnographic Practice*, The Lewis Henry Morgan Lectures. Chicago: University of Chicago Press.

Lave, J. 2019. *Learning and Everyday Life: Access, Participation, and Changing Practice*. Cambridge: Cambridge University Press.

Lave, J., and Rogoff, B. 1984. *Everyday Cognition: Its Development in Social Context*. Cambridge, MA: Harvard University Press.

Lave, J., and Wenger, E. 1991. *Situated Learning: Legitimate Peripheral Participation*. Cambridge: Cambridge University Press.

Law, J. 1987. "Technology and heterogeneous engineering: The case of Portuguese expansion." In W. E. Bijker, T. P. Hughes, and T. J. Pinch (eds.), *The Social Construction of Technological Systems: New Directions in the Sociology and History of Technology*. Cambridge, MA: MIT Press. 105–28.

Lawrence, S., and Davies, P. 2015. "Innovation, adaptation and technology as habitus: The origins of alluvial gold mining methods in Australia." *Archaeology in Oceania* 50(S1): 20–9. DOI: 10.1002/arco.5047.

Leary, J., ed. 2014. *Past Mobilities: Archaeological Approaches to Movement and Mobility*. London: Routledge.

Lechtman, H., and Merill, R. 1977. *Material Culture: Styles, Organization and the Dynamics of Technology*. St. Paul, MN: West Publishing.

Leclerc, M., Taché, K., Bedford, S., Spriggs, M., Lucquin, A., and Craig, O. E. 2018. "The use of Lapita pottery: Results from the first analysis of lipid residues." *Journal of Archaeological Science: Reports* 17: 712–22. DOI: 10.1016/j.jasrep.2017.12.019.

Leclerc, M., Grono, E., Bedford, S., and Spriggs, M. 2019a. "Assessment of the technological variability in decorated Lapita pottery from Teouma, Vanuatu, by petrography and LA-ICP-MS: Implications for Lapita social organisation." *Archaeological and Anthropological Sciences* 11(10): 5257–73. DOI: 10.1007/s12520-019-00862-z.

Leclerc, M., Taché, K., Bedford, S., and Spriggs, M. 2019b. "Organic residue analysis and the role of Lapita pottery." In M. Leclerc and J. Flexner (eds.), *Archaeologies of Island Melanesia*. Canberra: Australian National University Press. 179–90.

Lemonnier, P. 1986. "The study of material culture today: Toward an anthropology of technical systems." *Journal of Anthropological Archaeology* 5: 147–86.

Lemonnier, P. 1992. *Elements for an Anthropology of Technology*. Ann Arbor: University of Michigan Museum of Anthropology.

Lemonnier, P. 1993. "Introduction." In P. Lemonnier (ed.), *Technological Choices: Transformation in Material Cultures since the Neolithic*. London: Routledge. 1–35.

Lepore, J. 2014. "The disruption machine: What the gospel of innovation gets wrong." *The New Yorker*. June 23, 2014. www.newyorker.com/magazine/2014/06/23/the-disruption-machine.

Lepowsky, M. C. 1991. "The way of the ancestors: Custom, innovation, and resistance." *Ethnology* 30(3): 217–35. DOI: 10.2307/3773632.

Lestel, D., Brunois, F., and Gaunet, F. 2006. "Etho-ethnology and ethno-ethology." *Social Science Information* 45(2): 155–77. DOI: 10.1177/0539018406063633.

Leusch, V., Armbruster, B., Pernicka, E., and Slavčev, V. 2015. "On the invention of gold metallurgy: The gold objects from the Varna I cemetery (Bulgaria) – technological consequence and inventive creativity." *Cambridge Archaeological Journal* 25(1): 353–76. DOI: 10.1017/S0959774314001140.

Lévi-Strauss, C. 1996. *The Savage Mind*. London: Weidenfeld & Nicolson.

Lewis, H. M., and Laland, K. N. 2012. "Transmission fidelity is the key to the build-up of cumulative culture." *Philosophical Transactions of the Royal Society B: Biological Sciences* 367(1599): 2171–80. DOI: 10.1098/rstb.2012.0119.

Lewis, J., and Frieman, C. J. 2017. *A Geophysical Survey of Hall Rings Enclosure and Field System, Pelynt, Cornwall*. Southeast Kernow Archaeological Survey, Report No. 9.

Lewis, J., and Frieman, C. J. 2018. "Geophysical survey of four Cornish hillforts – later prehistoric settlements and gathering places in south-east Cornwall." *Cornish Archaeology* 57: 145–69.

Lienhard, J. H. 2006. *How Invention Begins: Echoes of Old Voices in the Rise of New Machines*. Oxford: Oxford University Press.

Liep, J. 2001. "Introduction." In J. Liep (ed.), *Locating Cultural Creativity*. London: Pluto Press. 1–13.

Lightfoot, K. G., and Martinez, A. 1995. "Frontiers and boundaries in archaeological perspective." *Annual Review of Anthropology* 24(1): 471–92. DOI: 10.1146/annurev.an.24.100195.002351.

Linares, O. F. 2002. "African rice (*Oryza glaberrima*): History and future potential." *Proceedings of the National Academy of Sciences* 99(25): 16360–5. DOI: 10.2307/3073959.

Linné, C. v. 1758. *Systema naturae per regna tria naturae, secundum classes, ordines, genera, species, cum characteribus, differentiis, synonymis, locis*. 10th edn., 2 vols. Stockholm: Impensis direct. Laurentii Salvii.

Linné, C. v. 1788. *Systema naturae per regna tria naturae, secundum classes, ordines, genera, species, cum characteribus, differentis, synonymis, locis*. Ed. G. E. Beer and J. F. Gmelin. 13th edn. 3 vols. Leipzig: Impensis Georg Emanuel Beer.

Lipson, M., Skoglund, P., Spriggs, M., *et al.* 2018. "Population turnover in Remote Oceania shortly after initial settlement." *Current Biology* 28(7): 1157–65.e7. DOI: 10.1016/j.cub.2018.02.051.

Lubart, T. I. 1998. "Creativity across cultures." In R. J. Sternberg (ed.), *Handbook of Creativity*. Cambridge: Cambridge University Press. 339–50.

Lubart, T. I. 2010. "Cross-cultural perspectives on creativity." In J. C. Kaufman and R. J. Sternberg (eds.), *The Cambridge Handbook of Creativity*. Cambridge: Cambridge University Press. 265–78.

Lycett, S. J., and Norton, C. J. 2010. "A demographic model for Palaeolithic technological evolution: The case of East Asia and the Movius Line." *Quaternary International* 211(1): 55–65. DOI: 10.1016/j.quaint.2008.12.001.

Lyman, R. L., and O'Brien, M. J. 1998. "The goals of evolutionary archaeology history and explanation." *Current Anthropology* 39(5): 615–52. DOI: 10.1086/204786.

McBrearty, S., and Brooks, A. S. 2000. "The revolution that wasn't: A new interpretation of the origin of modern human behavior." *Journal of Human Evolution* 39(5): 453–563. DOI: 10.1006/jhev.2000.0435.

McFall-Ngai, M., Hadfield, M. G., Bosch, T. C. G., *et al.* 2013. "Animals in a bacterial world, a new imperative for the life sciences." *Proceedings of the National Academy of Sciences* 110(9): 3229–36. DOI: 10.1073/pnas.1218525110.

McGovern, P. E., and Notis, M. D., eds. 1989. *Cross-Craft and Cross-Cultural Interactions in Ceramics*, Ceramics and Civilization 4. Westerville, OH: American Ceramic Society.

MacKenzie, D. A., and Wajcman, J. 1985. "Introductory essay." In D. A. MacKenzie and J. Wajcman (eds.), *The Social Shaping of Technology*. Milton Keynes: Open University Press. 2–25.

McKinnon, S. 2005a. *Neo-Liberal Genetics: The Myths and Moral Tales of Evolutionary Psychology*. Chicago: Prickly Paradigm Press.

McKinnon, S. 2005b. "On kinship and marriage: A critique of the genetic and gender calculus of evolutionary psychology." In S. McKinnon and S. Silverman (eds.), *Complexities: Beyond Nature and Nurture*. Chicago: University of Chicago Press. 106–31.

McLeod, J. 2018. "Who needs the Kardashians – meet the new breed of influencers tapping social media's 'gold rush.'" *Financial Post*. September 27, 2018. https://business.financialpost.com/technology/who-needs-the-kardashians-meet-the-new-breed-of-influencers-tapping-into-social-medias-advertising-gold-rush (accessed September 16, 2020).

McNiven, I. J. 2020. "Primordialising Aboriginal Australians: Colonialist tropes and Eurocentric views on behavioural markers of modern humans." In M. Porr and J. M. Matthews (eds.), *Interrogating Human Origins: Decolonisation and the Deep Human Past*. London: Routledge. 96–111.

McNiven, I. J., and Russell, L. 2005. *Appropriated Pasts: Indigenous Peoples and the Colonial Culture of Archaeology*. Oxford: AltaMira Press.

McNiven, I. J., David, B., Richards, T., *et al.* 2011. "New direction in human colonisation of the Pacific: Lapita settlement of south coast New Guinea." *Australian Archaeology* 72(1): 1–6. DOI: 10.1080/03122417.2011.11690525.

Malone, C., and Stoddart, S. 1998. "The conditions of creativity for prehistoric Maltese art." In S. J. Mithen (ed.), *Creativity in Human Evolution and Prehistory*. London: Routledge. 241–59.

Marchand, T. H. J. 2008. "Muscles, morals and mind: Craft apprenticeship and the formation of person." *British Journal of Educational Studies* 56(3): 245–71. DOI: 10.1111/j.1467-8527.2008.00407.x.

Marcuse, H. 1941. "Some social implications of modern technology." *Studies in Philosophy and Social Science* 9(3): 414–39. DOI: 10.4324/9780203208311-6.

Markides, C. 2006. "Disruptive innovation: In need of better theory." *Journal of Product Innovation Management* 23(1): 19–25. DOI: 10.1111/j.1540-5885.2005.00177.x.

Marks, J. 2020. "Naming the sacred ancestors: Taxonomic reification and Pleistocene genetic narratives." In M. Porr and J. M. Matthews (eds.), *Interrogating Human Origins: Decolonisation and the Deep Human Past*. London: Routledge. 295–309.

Marquis, C., and Tilcsik, A. 2013. "Imprinting: Toward a multilevel theory." *Academy of Management Annals* 7(1): 195–245. DOI: 10.5465/19416520.2013.766076.

Martin, R., and Simmie, J. 2008. "Path dependence and local innovation systems in city-regions." *Innovation* 10(2–3): 183–96. DOI: 10.5172/impp.453.10.2-3.183.

Mathieu, J. R., and Meyer, D. A. 1997. "Comparing axe heads of stone, bronze, and steel: Studies in experimental archaeology." *Journal of Field Archaeology* 24(3): 333–51. DOI: 10.1179/009346997792208122.

Mattingly, D. J. 2006. *An Imperial Possession: Britain in the Roman Empire, 54 BC–AD 409*, Penguin History of Britain. London: Allen Lane.

May, S. K. 2008. "Learning art, learning culture: Art, education, and the formation of new artistic identities in Arnhem Land, Australia." In I. D. Sanz, D. Fiore, and S. K. May (eds.), *Archaeologies of Art: Time, Place, Identity*. Walnut Creek, CA: Left Coast Press. 171–94.

May, S. K., Taçon, P. S. C., Wesley, D., and Travers, M. 2010. "Painting history: Indigenous observations and depictions of the 'other' in northwestern Arnhem Land, Australia." *Australian Archaeology* 71(1): 57–65. DOI: 10.1080/03122417.2010.11689384.

May, S. K., Taçon, P. S. C., Wesley, D., and Pearson, M. 2013. "Painted ships on a painted Arnhem Land landscape." *The Great Circle* 35(2): 83–102.

May, S. K., Wesley, D., Goldhahn, J., Litster, M., and Manera, B. 2017. "Symbols of power: The firearm paintings of Madjedbebe (Malakunanja II)." *International Journal of Historical Archaeology* 21(3): 690–707. DOI: 10.1007/s10761-017-0393-6.

Mazur, A. 1975. "Opposition to technological innovation." *Minerva* 13(1): 58–81.

Mellars, P. 1996. "The emergence of biologically modern populations in Europe: A social and cognitive 'revolution'?" In W. G. Runciman and J. Maynard (eds.), *Evolution of Social Behaviour Patterns in Primates and Man*, Proceedings of the British Academy 88. Oxford: Oxford University for the British Academy. 179–202.

Mellars, P. 2005. "The impossible coincidence. A single-species model for the origins of modern human behavior in Europe." *Evolutionary Anthropology: Issues, News, and Reviews* 14(1): 12–27. DOI: 10.1002/evan.20037.

Mellars, P. 2006. "Going east: New genetic and archaeological perspectives on the modern human colonization of Eurasia." *Science* 313(5788): 796. DOI: 10.1126/science.1128402.

Méry, S., Anderson, P., Inizan, M.-L., Lechevallier, M., and Pelegrin, J. 2006. "A pottery workshop with flint tools on blades knapped with copper at Nausharo (Indus civilisation, *ca.* 2500 BC)." *Journal of Archaeological Science* 34 (2007): 1098–116. DOI: 10.1016/j.jas.2006.10.002.

Meyer, A. D., and Goes, J. B. 1988. "Organizational assimilation of innovations: A multilevel contextual analysis." *Academy of Management Journal* 31(4): 897–923. DOI: 10.2307/256344.

Midgley, D. F., and Dowling, G. R. 1978. "Innovativeness: The concept and its measurement." *Journal of Consumer Research* 4(4): 229–42.

Migliano, A. B., Battiston, F., Viguier, S., *et al.* 2020. "Hunter-gatherer multilevel sociality accelerates cumulative cultural evolution." *Science Advances* 6(9). DOI: 10.1126/sciadv.aax5913.

Mills, B. J. 2018. "Intermarriage, technological diffusion, and boundary objects in the US Southwest." *Journal of Archaeological Method and Theory* 25(4): 1051–86. DOI: 10.1007/s10816-018-9392-0.

Mills, B. J., and Peeples, M. A. 2019. "Reframing diffusion through social network theory." In K. G. Harry and B. Roth (eds.), *Interaction and Connectivity in the Greater Southwest*. Boulder: University of Colorado Press. 40–62.

Mills, B. J., Clark, J. J., Peeples, M. A., *et al.* 2013. "Transformation of social networks in the late pre-Hispanic US Southwest." *Proceedings of the National Academy of Sciences* 110(15): 5785–90.

Mills, B. J., Clark, J. J., and Peeples, M. A. 2016. "Migration, skill, and the transformation of social networks in the pre-Hispanic Southwest." *Economic Anthropology* 3(2): 203–15. DOI: 10.1002/sea2.12060.

Mithen, S. J. 1996. *The Prehistory of the Mind: A Search for the Origins of Art, Science and Religion*. London: Thames and Hudson.

Mithen, S. J. 1998. "A creative explosion? Theory of mind, language and the disembodied mind of the Upper Palaeolithic." In S. J. Mithen (ed.), *Creativity in Human Evolution and Prehistory*. London: Routledge. 165–91.

Mokyr, J. 1990. *The Lever of Riches: Technological Creativity and Economic Progress*. Oxford: Oxford University Press.

Mokyr, J. 1992. "Technological inertia in economic history." *Journal of Economic History* 52(2): 325–38.

Montón-Subías, S., and Hernando-Gonzalo, A. 2018. "Modern colonialism, Eurocentrism and historical archaeology: Some engendered thoughts." *European Journal of Archaeology* 21(3): 455–71. DOI: 10.1017/eaa.2017.83.

Moore, G. A. 1991. *Crossing the Chasm: Marketing and Selling High-Tech Products to Mainstream Consumers*. 2nd edn. New York: Harper Business.

Moore, G. A. 1995. *Inside the Tornado: Marketing Strategies from Silicon Valley's Cutting Edge*. New York: Harper Business.

Morgan, L. H. 1985 [1877]. *Ancient Society*. Tucson: University of Arizona Press.

Morris, M. W., and Leung, K. 2010. "Creativity East and West: Perspectives and parallels." *Management and Organization Review* 6(3): 313–27. DOI: 10.1111/j.1740-8784.2010.00193.x.

Moser, P., and San, S. 2020. "Immigration, science, and invention: Lessons from the Quota Acts." https://ssrn.com/abstract=3558718. DOI: 10.2139/ssrn.3558718.

Moser, S. 1992. "The visual language of archaeology: A case study of the Neanderthals." *Antiquity* 66: 831–44. DOI: 10.1017/S0003598X0004477X.

Moser, S. 1993. "Gender stereotyping in pictorial reconstructions of human origins." In H. du Cros and L. Smith (eds.), *Women in Archaeology: A Feminist Critique*. Canberra: Australian National University Press. 75–92.

Moser, S. 1998. *Ancestral Images: The Iconography of Human Origins*. Stroud: Sutton.

Mouritsen, J., and Dechow, N. 2001. "Technologies of managing and the mobilization of paths." In R. Garud and P. Karnøe (eds.), *Path Dependence and Creation*. London: Lawrence Erlbaum. 355–79.

Mullen, D., and Birt, P. 2009. "Modernity and tradition: Considerations of Cornish identity in the archaeological record of a Burra dugout." *Australian Archaeology* 69: 59–67. DOI: 10.1080/03122417.2009.11681901.

Munn, N. D. 1992. "The cultural anthropology of time: A critical essay." *Annual Review of Anthropology* 21(1): 93–123. DOI: 10.1146/annurev.an.21.100192.000521.

Murphy, D. J. 2007. *People, Plants, and Genes: The Story of Crops and Humanity*. Oxford: Oxford University Press.

Murray, T., and Williamson, C. 2003. "Archaeology and history." In R. Manne (ed.), *Whitewash: On Keith Windschuttle's Fabrication of Aboriginal History*. Melbourne: Black Inc. Agenda. 311–33.

Nauright, J. 2005. "Cornish miners and the Witwatersrand gold mines in South Africa, c. 1890–1904." *Cornish History*. Available at http://www.marjon.ac.uk/cornish-history/witwatersrand/index.htm (accessed October 20, 2020).

NCC. 2013. "History programmes of study: Key stage 3." National curriculum in England Statutory guidance. London: Department of Education. https://assets.publishing.service.gov.uk/government/uploads/system/uploads/attachment_data/file/239075/SECONDARY_national_curriculum_-_History.pdf.

Neale, M. 2017. *Songlines: Tracking the Seven Sisters*. Canberra: National Museum of Australia Press.

Neff, H. 1992. "Ceramics and evolution." *Archaeological Method and Theory* 4: 141–93.

Neff, H. 1996. "Ceramics and evolution." In M. J. O' Brien (ed.), *Evolutionary Archaeology: Theory and Application*. Salt Lake City: University of Utah Press. 244–69.

Ng, A. K. 2001. *Why Asians Are Less Creative than Westerners*. London: Prentice Hall.

Nickens, P. R. 1991. "The destruction of archaeological sites and data." In G. Smith and J. Ehrenhard (eds.), *Protecting the Past*. Boca Raton, FL: CRC Press. 73–81.

Nielsen, F. 2001. "Why do farmers innovate and why don't they innovate more? Insights from a study in East Africa." In C. Reij and A. Waters-Bayer (eds.), *Farmer Innovation in Africa: A Source of Inspiration for Agricultural Development*. Sterling, VA: Earthscan. 92–103.

Nilsson, S. 1868. *The primitive inhabitants of Scandinavia. An essay on comparative ethnography, and a contribution to the history of the development of mankind: containing a description of the implements, dwellings, tombs, and mode of living of the savages in the north of Europe during the stone age*. 3rd edn. London: Longmans, Green.

Noble, S. U. 2018. *Algorithms of Oppression: How Search Engines Reinforce Racism*. New York: New York University Press.

Noury, A. 2019. "Along the roads of the Lapita people: Designs, groups and travels." In S. Bedford and M. Spriggs (eds.), *Debating Lapita: Distribution, Chronology, Society and Subsistence*. Canberra: Australian National University Press. 335–48.

Nowakowski, J. 2011. "Appraising the bigger picture – Cornish Iron Age and Romano-British lives and settlements 25 years on." *Cornish Archaeology* 50: 241–61.

Nowell, A. 2010. "Defining behavioral modernity in the context of Neandertal and anatomically modern human populations." *Annual Review of Anthropology* 39(1): 437–52. DOI: 10.1146/annurev.anthro.012809.105113.

Nunn, P. D., and Reid, N. J. 2016. "Aboriginal memories of inundation of the Australian coast dating from more than 7,000 years ago." *Australian Geographer* 47(1): 11–47. DOI: 10.1080/00049182.2015.1077539.

Nye, D. E. 2006. *Technology Matters: Questions to Live With*. Cambridge, MA: MIT Press.

O'Brien, M. J., and Shennan, S. J., eds. 2010. *Innovation in Cultural Systems: Contributions from Evolutionary Anthropology*. Cambridge, MA: MIT Press.

OECD and Statistical Office of European Communities. 2005. *Oslo Manual: Guidelines for Collecting and Interpreting Innovation Data*. Paris: OECD.

Okumura, M., and Araujo, A. G. M. 2014. "Long-term cultural stability in hunter-gatherers: A case study using traditional and geometric morphometric analysis of lithic stemmed bifacial points from Southern Brazil." *Journal of Archaeological Science* 45: 59–71. DOI: 10.1016/j.jas.2014.02.009.

Olausson, D. S. 2008. "Does practice make perfect? Craft expertise as a factor in aggrandizer strategies." *Journal of Archaeological Method and Theory* 15(1): 28–50. DOI: 10.1007/s10816-007-9049-x.

Olausson, D. S. 2017. "Knapping skill and craft specialization in Late Neolithic flint daggers." *Lithic Technology* 42(4): 127–39. DOI: 10.1017/S0003598X0004477X.

Osborne, R. 2008. "Introduction: For tradition as an analytical category." *World Archaeology* 40(3): 281–94. DOI: 10.1080/00438240802260806.

Ottaway, B. S. 1989. "Interactions of some of the earliest copper using cultures in Central Europe." In A. Hauptmann, E. Pernicka, and G. A. Wagner (eds.), *Archäometallugie der Alten Welt: Beiträge zum Internationalen Symposium "Old World Archaeometallurgy," Heidelberg 1987* = *Old World Archaeometallurgy: Proceedings of the International Symposium "Old World Archaeometallurgy," Heidelberg 1987*. Bochum: Selbstverlag des Deutschen Bergbau-Museums. 19–24.

Ottaway, B. S. 2001. "Innovation, production and specialisation in early prehistoric copper metallurgy." *European Journal of Archaeology* 4(1): 87–112. DOI: 10.1179/eja.2001.4.1.87.

Pálsson, G. 1994. "Enskilment at sea." *Man* 29(4): 901–27. DOI: 10.2307/3033974.

Paluck, E. L., Shepherd, H., and Aronow, P. M. 2016. "Changing climates of conflict: A social network experiment in 56 schools." *Proceedings of the National Academy of Sciences* 113(3): 566–71. DOI: 10.1073/pnas.1514483113.

Panich, L. M. 2013. "Archaeologies of persistence: Reconsidering the legacies of colonialism in Native North America." *American Antiquity* 78(1): 105–22. DOI: 10.7183/0002-7316.78.1.105.

Panich, L. M. 2020. *Narratives of Persistence: Indigenous Negotiations of Colonialism in Alta and Baja California*, Archaeology of Indigenous–Colonial Interactions in the Americas. Tucson: University of Arizona Press.

Papousek, D. A. 1989. "Technological change as social rebellion." In S. van der Leeuw and R. Torrence (eds.), *What's New? A Closer Look at the Process of Innovation*. London: Unwin Hywin. 140–66.

Parayil, G. 1991. "Schumpeter on invention, innovation and technological change." *Journal of the History of Economic Thought* 13(1): 78–89. DOI: 10.1017/S1053837200003412.

Park, S.-H. S., Lee, L., and Yi, M. Y. 2011. "Group-level effects of facilitating conditions on individual acceptance of information systems." *Information Technology and Management* 12(4): 315–34. DOI: 10.1007/s10799-011-0097-2.

Pascoe, B. 2016. *Dark Emu: Aboriginal Australia and the Birth of Agriculture*. London: Scribe.

Payton, P. 1999. *The Cornish Overseas*. Fowey: Alexander Associates.

Payton, P. 2001. "Cousin Jacks and ancient Britons: Cornish immigrants and ethnic identity." *Journal of Australian Studies* 25(68): 54–64. DOI: 10.1080/14443050109387662.

Payton, P. 2007. *Making Moonta: The Invention of Australia's Little Cornwall*. Exeter: University of Exeter Press.

Payton, P. 2014. "1848 and all that: Early South Australia and the Cornish radical tradition." *Journal of the Historical Society of South Australia* 42: 17–28.

Payton, P. 2016. *One and All: Labor and the Radical Tradition in South Australia*. Adelaide: Wakefield Press.

Pelegrin, J. 1990. "Prehistoric lithic technology: Some aspects of research." *Archaeological Review from Cambridge* 9(1): 116–25.

Peregrine, P. N. 1996. "Archaeology and world-systems theory." *Sociological Inquiry* 66(4): 486–95. DOI: 10.1111/j.1475–682X.1996.tb01189.x.

Pétrequin, P., Cassen, S., Errera, M., Klassen, L., Sheridan, A., and Pétrequin, A.-M., eds. 2012. *JADE: Grandes haches alpines du Néolithique européen, Ve au IVe millénaires av. J.-C.* Besançon: Presses Universitaires de Franche-Comté.

Petrov, A. N. 2011. "Beyond spillovers: Interrogating innovation and creativity in the peripheries." In H. Bathelt, M. Feldman, and D. F. Kogler (eds.), *Beyond Territory: Dynamic Geographies of Innovation and Knowledge Creation*. London: Routledge. 168–90.

Petrov, A. N. 2016. "Exploring the Arctic's 'other economies': Knowledge, creativity and the new frontier." *Polar Journal* 6(1): 51–68. DOI: 10.1080/2154896X.2016.1171007.

Pfaffenberger, B. 1988. "Fetishised objects and humanised nature: Towards an anthropology of technology." *Man* 23(2): 236–52. DOI: 10.2307/2802804.

Pfaffenberger, B. 1992. "Social anthropology of technology." *Annual Review of Anthropology* 21: 491–516. DOI: 10.2307/2155997.

Pfeiffer, J. E. 1982. *The Creative Explosion: An Inquiry into the Origins of Art and Religion*. New York: Harper & Row.

Piezonka, H. 2012. "Stone Age hunter-gatherer ceramics of north-eastern Europe: New insights into the dispersal of an essential innovation." *Documenta praehistorica* 39: 22–51. DOI: 10.4312/dp.39.2.

Pinch, T. J., and Bijker, W. E. 1987. "The social construction of facts and artifacts; or, How the sociology of science and the sociology of technology might benefit each other." In W. E. Bijker, T. P. Hughes, and T. J. Pinch (eds.), *The Social Construction of Technological Systems: New Directions in the Sociology and History of Technology*. Cambridge, MA: MIT Press. 17–50.

Piperno, D. R. 2012. "New archaeobotanical information on early cultivation and plant domestication involving microplant (phytolith and starch grain) remains." In P. Gepts, T. R. Famula, R. L. Bettinger, S. B. Brush, A. B. Damania, P. E. McGuire, and C. O. Qualset (eds.), *Biodiversity in Agriculture: Domestication, Evolution, and Sustainability*. Cambridge: Cambridge University Press. 136–59.

Porr, M., and Matthews, J. M. 2017. "Post-colonialism, human origins and the paradox of modernity." *Antiquity* 91(358): 1058–68. DOI: 10.15184/aqy.2017.82.

Porr, M., and Matthews, J. M. 2020. "Interrogating and decolonising the deep human past." In M. Porr and J. M. Matthews (eds.), *Interrogating Human Origins: Decolonisation and the Deep Human Past*. London: Routledge. 3–31.

Porter, B. 2010. "Testing the limits of nonzero: Cooperation, conflict and hierarchy in ancient Near Eastern marginal environments." In R. C. Marshall (ed.), *Cooperation in Economy and Society*. Lanham, MD: Altamira Press. 149–74.

Posth, C., Nägele, K., Colleran, H., *et al.* 2018. "Language continuity despite population replacement in Remote Oceania." *Nature Ecology & Evolution* 2(4): 731–40. DOI: 10.1038/s41559-018-0498-2.

Powell, A., Shennan, S., and Thomas, M. G. 2009. "Late Pleistocene demography and the appearance of modern human behavior." *Science* 324(5932): 1298. DOI: 10.1126/science.1170165.

Premo, L. S., and Kuhn, S. L. 2010. "Modeling effects of local extinctions on culture change and diversity in the Paleolithic." *PLOS ONE* 5(12). DOI: 10.1371/journal.pone.0015582.

Prentiss, A. M., ed. 2019. *Handbook of Evolutionary Research in Archaeology.* Cham: Springer.

Prentiss, A. M., and Laue, C. L. 2019. "Cultural macroevolution." In A. M. Prentiss (ed.), *Handbook of Evolutionary Research in Archaeology.* Cham: Springer. 111–26.

Prentiss, A. M., and Lenart, M. 2009. "Cultural stasis and change in northern North America: A macroevolutionary perspective." In A. M. Prentiss, I. Kuijt, and J. C. Chatters (eds.), *Macroevolution in Human Prehistory: Evolutionary Theory and Processual Archaeology.* London: Springer. 235–52.

Price, L. L., and Ridgway, N. M. 1983. "Development of a scale to measure use innovativeness." In R. P. Bagozzi and A. M. Tybout (eds.), *Advances in Consumer Research.* Ann Arbor: Association for Consumer Research. 679–84.

Price, T. E. 2009. "Ancient farming in eastern North America." *Proceedings of the National Academy of Sciences* 106(16): 6427. DOI: 20.1073/pnas.0902617106.

Price, T. D., and Bar-Yosef, O. 2011. "The origins of agriculture: New data, new ideas. An introduction to Supplement 4. *Current Anthropology* 52(S4): S163–74. DOI: 10.1086/659964.

Price, T. D., and Feinman, G. M., eds. 2010. *Pathways to Power: New Perspectives on the Emergence of Social Inequality,* Fundamental Issues in Archaeology. New York: Springer.

Pryor, O. 1969. *Australia's Little Cornwall.* Moonta: National Trust, Moonta Branch.

Pulak, C. 2012. "Uluburun shipwreck." In E. H. Cline (ed.), *The Oxford Handbook of the Bronze Age Aegean.* Oxford: Oxford University Press. 862–76.

Quinnell, H. 1987. "Cornish gabbroic pottery: The development of a hypothesis." *Cornish Archaeology* 26: 7–12.

Quinnell, H. 2004. *Trethurgy. Excavations at Trethurgy Round, St. Austell: Community and Status in Roman and Post-Roman Cornwall.* Truro: Cornwall County Council.

Rabey, M. A. 1989. "Technological continuity and change among the Andean peasants: Opposition between local and global strategies." In S. E. van der Leeuw and R. Torrence (eds.), *What's New? A Closer Look at the Process of Innovation.* London: Unwin Hyman. 167–81.

Radivojević, M. 2015. "Inventing metallurgy in western Eurasia: A look through the microscope lens." *Cambridge Archaeological Journal* 25(1): 321–38. DOI: 10.1017/S0959774314001097.

Radivojević, M., and Rehren, T. 2016. "Paint it black: The rise of metallurgy in the Balkans." *Journal of Archaeological Method and Theory* 23(1): 200–37. DOI: 10.1007/s10816-014-9238-3.

Radivojević, M., Rehren, T., Pernicka, E., Šljivar, D., Brauns, M., and Borić, D. 2010. "On the origins of extractive metallurgy: New evidence from Europe." *Journal of Archaeological Science* 37(11): 2775–87. DOI: 10.1016/j.jas.2010.06.012.

Radivojević, M., Rehren, T., Kuzmanović-Cvetković, J., Jovanović, M., and Northover, J. P. 2013. "Tainted ores and the rise of tin bronzes in Eurasia, c. 6,500 years ago." *Antiquity* 87(338): 1030–45. DOI: 10.1017/S0003598X0004984X.

Ram, S., and Jung, H.-S. 1994. "Innovativeness in product usage: A comparison of early adopters and early majority." *Psychology & Marketing* 11(1): 57–67. DOI: 10.1002/mar.4220110107.

Randall, A. 1995. "Reinterpreting 'Luddism': Resistance to new technology in the British Industrial Revolution." In M. Bauer (ed.), *Resistance to New Technology: Nuclear Power, Information Technology and Biotechnology.* Cambridge: Cambridge University Press. 57–80.

Randsborg, K. 1998. "Plundered Bronze Age graves: Archaeological and social implications." *Acta archaeologica* 69: 113–38.

Read, D. 2006. "Tasmanian knowledge and skill: Maladaptive imitation or adequate technology? *American Antiquity* 71(1): 164–84. DOI: 10.2307/40035327.

Redding, R. W. 1988. "A general explanation of subsistence change: From hunting and gathering to food production." *Journal of Anthropological Archaeology* 7(1): 56–97. DOI: 10.1016/0278-4165(88)90007-4.

Rehn, A., and De Cock, C. 2009. "Deconstructing creativity." In T. Rickards, M. A. Runco, and S. Moger (eds.), *The Routledge Companion to Creativity.* London: Routledge. 222–31.

Rehn, A., and Vachhani, S. 2006. "Innovation and the post-original: On moral stances and reproduction." *Creativity and Innovation Management* 15(3): 310–22. DOI: 10.1111/j.1467-8691.2006.00390.x.

Renfrew, C. 1967. "Colonialism and megalithismus." *Antiquity* 41(164): 276–88. DOI: 10.1017/S0003598X00033512.

Renfrew, C. 1970. "The autonomy of the south-east European Copper Age." *Proceedings of the Prehistoric Society* 35: 12–47. DOI: 10.1017/S0079497X00013396.

Renfrew, C. 1973. "Wessex as a social question." *Antiquity* 47 (187): 221–5. DOI: 10.1017/S0003598X00103928.

Renfrew, C. 1986. "Introduction: Peer–polity interaction and socio-political change." In C. Renfrew and J. F. Cherry (eds.), *Peer–Polity Interaction and Socio-Political Change.* Cambridge: Cambridge University Press. 1–18.

Renfrew, C. 2009. "Situating the creative explosion: Universal or local?" In C. Renfrew and I. Morley (eds.), *Becoming Human: Innovation in Prehistoric Material and Spiritual Culture.* Cambridge: Cambridge University Press. 74–94.

Richerson, P. J., and Boyd, R. 2005. *Not by Genes Alone: How Culture Transformed Human Evolution.* Chicago: University of Chicago Press.

Richerson, P. J., Boyd, R., and Bettinger, R. L. 2009. "Cultural innovations and demographic change." *Human Biology* 81(3): 211–35. DOI: 10.2307/41466599.

Rickards, T., Runco, M. A., and Moger, S. 2009. *The Routledge Companion to Creativity.* London: Routledge.

Ridgway, N. M., and Price, L. L. 1994. "Exploration in product usage: A model of use innovativeness." *Psychology & Marketing* 11(1): 69–84. DOI: 10.1002/mar.4220110108.

Riede, F. 2006. "The Scandinavian connection: The roots of Darwinian archaeology in 19th-century Scandinavian archaeology." *Bulletin of the History of Archaeology* 16(1): 4–19. DOI: 10.5334/bha.16102.

Riede, F., Johannsen, N. N., Högberg, A., Nowell, A., and Lombard, M. 2018. "The role of play objects and object play in human cognitive evolution and innovation." *Evolutionary Anthropology: Issues, News, and Reviews* 27(1): 46–59. DOI: 10.1002/evan.21555.

Rindos, D., Aschmann, H., Bellwood, P., *et al.* 1980. "Symbiosis, instability, and the origins and spread of agriculture: A new model [and comments and reply]." *Current Anthropology* 21(6): 751–72. DOI: 10.2307/2742515.

Robb, J., and Pauketat, T. R. 2013. "From moments to millennia: Theorizing scale and change in human history." In T. R. Pauketat and J. Robb (eds.), *Big Histories, Human Lives: Tackling Problems of Scale in Archaeology*. Santa Fe: School for Advanced Research Press. 3–33.

Roberts, B. W., and Frieman, C. J. 2012. "Drawing boundaries and building models: Investigating the concept of the 'Chalcolithic frontier' in northwest Europe." In M. J. Allen, J. Gardiner, and J. A. Sheridan (eds.), *Is There a British Chalcolithic? People, Place and Polity in the Later 3rd Millennium*, Prehistoric Society Research Paper 4. Oxford: The Prehistoric Society and Oxbow Books. 27–39.

Roberts, B. W., and Frieman, C. J. 2015. "Early metallurgy in western and northern Europe." In C. Fowler, J. Harding, and D. Hofmann (eds.), *The Oxford Handbook of Neolithic Europe*. Oxford: Oxford University Press. 711–28.

Roberts, B. W., and Radivojević, M. 2015. "Invention as a process: Pyrotechnologies in early societies." *Cambridge Archaeological Journal* 25(1): 299–306. DOI: 10.1017/S0959774314001188.

Roberts, B. W., Thornton, C., and Pigott, V. C. 2009. "Development of metallurgy in Eurasia." *Antiquity* 83: 1012–22. DOI: 10.1017/S0003598X00099312.

Robins, K., and Webster, F. 1999. *Times of the Technoculture: From the Information Society to the Virtual Life*. London: Routledge.

Roddick, A. P., and Stahl, A. B., eds. 2016. *Knowledge in Motion: Constellations of Learning across Time and Place*. Tucson: University of Arizona Press.

Rodríguez-Vidal, J., d'Errico, F., Pacheco, F. G., *et al.* 2014. "A rock engraving made by Neanderthals in Gibraltar." *Proceedings of the National Academy of Sciences* 111(37): 13301–6. DOI: 10.1073/pnas.1411529111.

Rogers, E. M. 1962. *Diffusion of Innovations*. New York: Free Press of Glencoe.

Rogers, E. M. 2003. *Diffusion of Innovations*. 5th edn. London: Free Press.

Rogers, E. M., and Shoemaker, F. F. 1971. *Communication of Innovations: A Cross-Cultural Approach*. 2nd edn. New York: Free Press.

Rosaldo, R., Lavie, S., and Narayan, K. 1993. "Introduction: Creativity in anthropology." In S. Lavie, K. Narayan, and R. Rosaldo (eds.), *Creativity/Anthropology*. Ithaca, NY: Cornell University Press. 1–8.

Rose, P., and Preston-Jones, A. 1995. "Changes in the Cornish countryside AD 400–1100." In D. Hooke and S. Burnell (eds.), *Landscape and Settlement in Britain AD 400–1066*. Exeter: University of Exeter Press. 51–68.

Rosenthal, C. 2018. *Accounting for Slavery: Masters and Management*. Cambridge, MA: Harvard University Press.

Roth, H. L. 1899. *The Aborigines of Tasmania*. 2nd edn. Halifax: F. King & Sons.

Roud, S. 2011. *The Lore of the Playground: The Children's World – Then and Now.* London: Arrow.

Roux, V. 2003. "A dynamic systems framework for studying technological change: Application to the emergence of the potter's wheel in the southern Levant." *Journal of Archaeological Method and Theory* 10(1): 1–30. DOI: 10.2307/20177471.

Roux, V. 2010. "Technological innovation and developmental trajectories: Social factors as evolutionary forces." In M. J. O'Brien and S. J. Shennan (eds.), *Innovation in Cultural Systems: Contributions from Evolutionary Anthropology.* Cambridge, MA: MIT Press. 217–34.

Rowley-Conwy, P. 2007. *From Genesis to Prehistory: The Archaeological Three Age System and Its Contested Reception in Denmark, Britain, and Ireland,* Oxford Studies in the History of Archaeology. Oxford: Oxford University Press.

Rudowicz, E. 2003. "Creativity and culture: A two way interaction." *Scandinavian Journal of Educational Research* 47(3): 273–90. DOI: 10.1080/00313830308602.

Ryan, B., and Gross, N. C. 1943. "The diffusion of hybrid seed corn in two Iowa communities." *Rural Sociology* 8(1): 15–24.

Ryan, B., and Gross, N. C. 1950. "Acceptance and diffusion of hybrid corn seed in two Iowa communities." *Research Bulletin (Iowa Agriculture and Home Economics Experiment Station)* 29(372): 663–708.

Sabatini, S., Bergerbrant, S., Brandt, L. Ø., Margaryan, A., and Allentoft, M. E. 2019. "Approaching sheep herds origins and the emergence of the wool economy in continental Europe during the Bronze Age." *Archaeological and Anthropological Sciences.* DOI: 10.1007/s12520-019-00856-x.

Sahlins, M. 2013. "On the culture of material value and the cosmography of riches." *HAU: Journal of Ethnographic Theory* 3(2): 161–95. DOI: 10.14318/hau3.2.010.

Sahlins, M., and Service, E. R. 1960. *Evolution and Culture.* Ann Arbor: University of Michigan Press.

Sánchez-Quinto, F., Malmström, H., Fraser, M., *et al.* 2019. "Megalithic tombs in western and northern Neolithic Europe were linked to a kindred society." *Proceedings of the National Academy of Sciences* 116(19). DOI: 10.1073/pnas.1818037116.

Sand, C. 2007. "Looking at the big motifs: A typology of the central band decorations of the Lapita ceramic tradition of New Caledonia (Southern Melanesia) and preliminary regional comparisons." In C. Sand, S. Bedford, and S. P. Connaughton (eds.), *Oceanic Explorations.* Canberra: Australian National University Press. 265–88.

Sapsed, J., and Tschang, F. T. 2014. "Art is long, innovation is short: Lessons from the Renaissance and the digital age." *Technological Forecasting and Social Change* 83: 127–41. DOI: 10.1016/j.techfore.2013.09.014.

Saraydar, S., and Shimada, I. 1971. "A quantitative comparison of efficiency between a stone axe and a steel axe." *American Antiquity* 36(2): 216–7. DOI: 10.2307/278680.

Sayes, E. 2014. "Actor-Network Theory and methodology: Just what does it mean to say that nonhumans have agency?" *Social Studies of Science* 44(1): 134–49. DOI: 10.1177/0306312713511867.

Scharl, S. 2016. "Patterns of innovation transfer and the spread of copper metallurgy to Central Europe." *European Journal of Archaeology* 19(2): 215–44. DOI: 10.1080/14619571.2016.1147313.

Schiffer, M. B. 1983. "Toward the identification of formation processes." *American Antiquity* 48: 675–706.

Schiffer, M. B. 1987. *Formation Processes of the Archaeological Record.* Albuquerque: University of New Mexico Press.

Schiffer, M. B. 1996. "Some relationships between behavioral and evolutionary archaeologies." *American Antiquity* 61(4): 643–62. DOI: 10.2307/282009.

Schiffer, M. B. 2011. *Studying Technological Change: A Behavioral Approach,* Foundations of Archaeological Inquiry. Salt Lake City: University of Utah Press.

Schiffer, M. B., and Skibo, J. M. 1987. "Theory and experiment in the study of technological change." *Current Anthropology* 28(5): 595–622. DOI: 10.1086/203601.

Schlechty, P. C. 2001. *Shaking Up the Schoolhouse: How to Support and Sustain Educational Innovation.* San Francisco: Jossey-Bass.

Schnapp, A. 1996. *The Discovery of the Past: The Origins of Archaeology.* London: British Museum Press.

Schraube, E. 2009. "Technology as materialized action and its ambivalences." *Theory & Psychology* 19(2): 296–312. DOI: 10.1177/0959354309103543.

Schulting, R., and Richards, C. 2002. "The wet, the wild and the domesticated: The Mesolithic–Neolithic transition on the west coast of Scotland." *European Journal of Archaeology* 5(2): 147–89. DOI: 10.1177/14619571020050020201.

Schumpeter, J. A. 1934. *The Theory of Economic Development: An Inquiry into Profits, Capital, Credit, Interest, and the Business Cycle,* Harvard Economic Studies. Cambridge, MA: Harvard University Press.

Schumpeter, J. A. 1939. *Business Cycles: A Theoretical, Historical, and Statistical Analysis of the Capitalist Process.* 2 vols. New York: McGraw-Hill.

Schumpeter, J. A. 1943. *Capitalism, Socialism, and Democracy.* London: G. Allen & Unwin.

Schuster, C. E. 2015. *Social Collateral: Women and Microfinance in Paraguay's Smuggling Economy.* Berkeley: University of California Press.

Shane, S. A. 1992. "Why do some societies invent more than others?" *Journal of Business Venturing* 7(1): 29–46. DOI: 10.1016/0883-9026(92)90033-N.

Shanks, M., and Tilley, C. Y. 1987. *Re-Constructing Archaeology: Theory and Practice,* New Studies in Archaeology. Cambridge: Cambridge University Press.

Sharp, L. 1952. "Steel axes for Stone Age Australians." In E. H. Spicer (ed.), *Human Problems in Technological Change: A Casebook.* New York: Russel Sage Foundation. 69–90.

Shearmur, R. 2012. "Are cities the font of innovation? A critical review of the literature on cities and innovation." *Cities* 29: S9–S18. DOI: 10.1016/j.cities.2012.06.008.

Shearmur, R. 2015. "Far from the madding crowd: Slow innovators, information value, and the geography of innovation." *Growth and Change* 46(3): 424–42. DOI: 10.1111/grow.12097.

Shennan, S. 1989. "Cultural transmission and cultural change." In S. E. van der Leeuw and R. Torrence (eds.), *What's New? A Closer Look at the Process of Innovation.* London: Unwin Hyman. 330–46.

Shennan, S. 2000. "Population, culture history, and the dynamics of culture change." *Current Anthropology* 41(5): 811–35. DOI: 10.1086/317403.

Shennan, S. 2001. "Demography and cultural innovation: A model and its implications for the emergence of modern human culture." *Cambridge Archaeological Journal* 11(1): 5–16. DOI: 10.1017/S0959774301000014.

Shennan, S. 2002a. *Genes, Memes and Human History: Darwinian Archaeology and Cultural Evolution.* London: Thames and Hudson.

Shennan, S. 2002b. "Learning." In J. P. Hart and J. E. Terrell (eds.), *Darwin and Archaeology: A Handbook of Key Concepts.* Westport, CT: Bergen and Garvey. 183–99.

Shennan, S., and Steele, J. 1999. "Cultural learning in hominids: A behavioural ecological approach." In H. O. Box and K. R. Gibson (eds.), *Mammalian Social Learning: Comparative and Ecological Perspectives.* Cambridge: Cambridge University Press. 367–88.

Shennan, S. J., and Wilkinson, J. R. 2001. "Ceramic style change and neutral evolution: A case study from Neolithic Europe." *American Antiquity* 66(4): 577–93. DOI: 10.2307/2694174.

Sherratt, A. 1993. "What would a Bronze Age world-system look like? Relations between temperate Europe and the Mediterranean in later prehistory." *Journal of European Archaeology* 1(2): 1–58. DOI: 10.1179/096576693800719293.

Sherratt, A., and Taylor, T. 1997. "Metal vessels in Bronze Age Europe and the context of Vulchetrun." In A. Sherratt (ed.), *Economy and Society in Prehistoric Europe: Changing Perspectives.* Princeton: Princeton University Press. 431–56.

Shih, C.-F., and Venkatesh, A. 2004. "Beyond adoption: Development and application of a use-diffusion model." *Journal of Marketing* 68(1): 59–72. DOI: 10.1509/jmkg.68.1.59.24029.

Shimada, I., ed. 2007. *Craft Production in Complex Societies: Multicraft and Producer Perspectives.* Salt Lake City: University of Utah Press.

Shipton, C., Blinkhorn, J., Breeze, P. S., *al.* 2018. "Acheulean technology and landscape use at Dawadmi, central Arabia." *PLOS ONE* 13(7). DOI: 10.1371/journal.pone.0200497.

Shirky, C. 2015. "China's version of the 'Maker Movement' puts the US to shame." *Fortune.* http://fortune.com/2015/10/21/chinas-version-of-the-maker-movment-puts-the-u-s-to-shame/ (accessed September 18, 2020).

Silva, F., Stevens, C. J., Weisskopf, A., *et al.* 2016. "Modelling the geographical origin of rice cultivation in Asia using the Rice Archaeological Database." *PLOS ONE* 10(9). DOI: 10.1971/journal.pone.0137024.

Sim, R. 1999. "Why the Tasmanians stopped eating fish: Evidence for the late Holocene expansion in resource exploitation strategies." In J. Hall and I. J. McNiven (eds.), *Australian Coastal Archaeology.* Canberra: Australian National University Press. 263–9.

Singh, S. 2006. "Cultural differences in, and influences on, consumers' propensity to adopt innovations." *International Marketing Review* 23(2): 173–91.

Singleton, J. 1989. "Japanese folkcraft pottery apprenticeship: Cultural patterns of an educational institution." In M. W. Coy (ed.), *Apprenticeship: From Theory to Method and Back Again.* Albany: State University of New York Press. 13–30.

Smith, B. D. 2001. "Low-level food production." *Journal of Archaeological Research* 9(1): 1–43. DOI: 10.2307/41053172.

Smith, B. D. 2016. "Neo-Darwinism, niche construction theory, and the initial domestication of plants and animals." *Evolutionary Ecology* 30(2): 307–24. DOI: 10.1007/s10682-015-9797-0.

Smith, E. 2019. "Apprenticeships and 'future work': Are we ready?" *International Journal of Training and Development* 23(1): 69–88. DOI: 10.1111/ijtd.12145.

Smith, L. 2004. *Archaeological Theory and the Politics of Cultural Heritage.* London: Routledge.

Snell, D. C. 1997. *Life in the Ancient Near East, 3100–332 BCE.* New Haven: Yale University Press.

Sofaer, J. R. 2015. *Clay in the Age of Bronze: Essays in the Archaeology of Prehistoric Creativity.* New York: Cambridge University Press.

Sofaer, J. R., ed. 2018. *Considering Creativity: Creativity, Knowledge and Practice in Bronze Age Europe.* Oxford: Archaeopress Archaeology.

Sørensen, M. L. S. 1989. "Ignoring innovation – denying change: The role of iron and the impact of external influences on the transformation of Scandinavian societies 800–500 BC." In S. E. van der Leeuw and R. Torrence (eds.), *What's New? A Closer Look at the Process of Innovation.* London: Unwin Hyman. 182–202.

Specht, J., Denham, T., Goff, J., and Terrell, J. E. 2014. "Deconstructing the Lapita cultural complex in the Bismarck Archipelago." *Journal of Archaeological Research* 22(2): 89–140. DOI: 10.1007/s10814-013-9070-4.

Spriggs, M. 1997. *The Island Melanesians.* Oxford: Blackwell.

Spriggs, M. 2006. "The Lapita culture and Austronesian prehistory in Oceania." In P. Bellwood, J. J. Fox, and D. Tryon (eds.), *The Austronesians: Historical and Comparative Perspectives.* Canberra: Australian National University Press. 119–42.

Stanko, M. A. 2016. "Toward a theory of remixing in online innovation communities." *Information Systems Research* 27(4): 773–91. DOI: 10.1287/isre.2016.0650.

Star, S. L. 1989. "The structure of ill-structured solutions: Boundary objects and heterogeneous distributed problem solving." In L. Gasser and M. N. Huhns (eds.), *Distributed Artificial Intelligence.* London: Pitman. 37–54.

Star, S. L. 2010. "This is not a boundary object: Reflections on the origin of a concept." *Science, Technology & Human Values* 35(5): 601–17. DOI: 10.1177/0162243910377624.

Star, S. L., and Griesemer, J. R. 1989. "Institutional ecology, 'translations' and boundary objects: Amateurs and professionals in Berkeley's Museum of Vertebrate Zoology, 1907–1939." *Social Studies of Science* 19(3): 387–420. DOI: 10.2307/285080.

Stark, M. 1991. "Production and community specialization: A Kalinga ethnoarchaeological study." *World Archaeology* 23(1): 64–78. DOI: 10.1080/00438243.1991.9980159.

Stein, G. J. 1999. "Rethinking world-systems: Power, distance, and diasporas in the dynamics of interregional interaction." In P. N. Kardulias (ed.), *World-Systems Theory in Practice: Leadership, Production, and Exchange.* Oxford: Rowman & Littlefield. 153–77.

Steiniger, D. 2015. "On flint and copper daggers in Chalcolithic Italy." In C. J. Frieman and B. V. Eriksen (eds.), *Flint Daggers In Prehistoric Europe and Beyond*. Oxford: Oxbow. 45–56.

Stiner, M. C., and Kuhn, S. L. 2006. "Changes in the 'connectedness' and resilience of Paleolithic societies in Mediterranean ecosystems." *Human Ecology* 34(5): 693–712. DOI: 10.1007/s10745-006-9041-1.

Stockton, J. 1982. "Stone wall fish-traps in Tasmania." *Australian Archaeology* 14: 107–14. DOI: 10.2307/40286414.

Strathern, M. 1988. *The Gender of the Gift: Problems with Women and Problems with Society in Melanesia*, Studies in Melanesian Anthropology 6. Berkeley: University of California Press.

Strathern, M. 1996. "Cutting the network." *Journal of the Royal Anthropological Institute* 2(3): 517–35. DOI: 10.2307/3034901.

Strathern, M. 2004. "Transactions: An analytical foray." In E. Hirsch and M. Strathern (eds.), *Transactions and Creations: Property Debates and the Stimulus of Melanesia*. New York: Berghahn. 85–109.

Subedi, A., and Garforth, C. 1996. "Gender, information and communication networks: Implications for extension." *European Journal of Agricultural Education and Extension* 3(2): 63–74. DOI: 10.1080/13892249685300201.

Summerhayes, G. R. 2001. "Lapita in the far west: Recent developments." *Archaeology in Oceania* 36(2): 53–63. DOI: 10.1002/j.1834-4453.2001.tb00478.x.

Summerhayes, G. R. 2007. "The rise and transformation of Lapita in the Bismark archipelago." In S. Chiu and C. Sand (eds.), *From Southeast Asia to the Pacific: Archaeological Perspectives on the Austronesian Expansion and the Lapita Cultural Complex*. Taipei: Centre for Archaeological Studies, Academia Sinica. 141–69.

Sutton-Smith, B., Mechling, J., Johnson, T. W., and McMahon, F. R., eds. 1999. *Children's Folklore: A Source Book*. Logan: Utah State University Press.

Taçon, P. 2005. "Chains of connection." *Griffith Review* 9.

Taçon, P. S. C., and May, S. K. 2013. "Rock art evidence for Macassan–Aboriginal contact in northwestern Arnhem Land." In M. Clark and S. K. May (eds.), *Macassan History and Heritage: Journeys, Encounters and Influences*. Canberra: Australian National University Press. 127–39.

Taçon, P. S. C., May, S. K., Fallon, S. J., Travers, M., Wesley, D., and Lamilami, R. 2010. "A minimum age for early depictions of Southeast Asian praus in the rock art of Arnhem Land, Northern Territory." *Australian Archaeology* 71(1): 1–10. DOI: 10.1080/03122417.2010.11689379.

Taçon, P. S. C., Paterson, A., Ross, J., and May, S. K. 2012. "Picturing change and changing pictures: Contact period rock art of Australia." In J. McDonald and P. Veth (eds.), *A Companion to Rock Art*. Chichester: Blackwell. 420–36.

Tagliamonte, S. A., and D'Arcy, A. 2009. "Peaks beyond phonology: Adolescence, incrementation, and language change." *Language* 85(1): 58–108. DOI: 10.2307/40492846.

Taylor, R. 2007. "The polemics of eating fish in Tasmania: The historical evidence revisited." *Aboriginal History* 31: 1–26. DOI: 10.2307/24046726.

Taylor, R. 2008. "The polemics of making fire in Tasmania: The historical evidence revisited." *Aboriginal History* 32: 1–26. DOI: 10.2307/24046786.

Taylor, T. 1999. "Envaluing metal: Theorizing the Eneolithic 'hiatus.'" In S. M. M. Young, A. M. Pollard, P. Budd, and R. A. Ixer (eds.), *Metals in Antiquity*. Oxford: BAR. 22–32.

Taylor, T. 2010. *The Artificial Ape: How Technology Changed the Course of Human Evolution*. New York: Palgrave Macmillan.

Tehrani, J. J., and Riede, F. 2008. "Towards an archaeology of pedagogy: Learning, teaching and the generation of material culture traditions." *World Archaeology* 40(3): 316–31. DOI: 10.2307/40388216.

Tennie, C., Call, J., and Tomasello, M. 2009. "Ratcheting up the ratchet: On the evolution of cumulative culture." *Philosophical Transactions of the Royal Society B: Biological Sciences* 364(1528): 2405–15. DOI: 10.1098/rstb.2009.0052.

Terrell, J. E., Hunt, T. L., and Gosden, C. 1997. "The dimensions of social life in the Pacific: Human diversity and the myth of the primitive isolate." *Current Anthropology* 38(2): 155–95. DOI: 10.1086/204604.

Thomas, A. 1995. "The house the kids built: The gay black imprint on American dance music." In A. Doty and C. K. Creekmur (eds.), *Out in Culture: Gay, Lesbian, and Queer Essays on Popular Culture*. Durham, NC: Duke University Press. 437–63.

Thomas, H., Becerra, L., and Garrido, S. 2017. "Socio-technical dynamics of counter-hegemony and resistance." In B. Godin and D. Vinck (eds.), *Critical Studies of Innovation: Alternative Approaches to the Pro-Innovation Bias*. Cheltenham: Edward Elgar. 182–200.

Thomas, K. 2016. *Regulating Style: Intellectual Property Law and the Business of Fashion in Guatemala*. Berkeley: University of California Press.

Thorpe, C. M. 2011. "The early medieval native pottery of Cornwall: AD 400–1066." In S. Pearce (ed.), *Recent Archaeological Work in South-Western Britain: Papers in Honour of Henrietta Quinnell*. Oxford: BAR. 151–8.

Tilley, C. Y. 1989. "Archaeology as socio-political action in the present." In V. Pinsky and A. Wylie (eds.), *Critical Traditions in Contemporary Archaeology: Essays in the Philosophy, History and Socio-Politics of Archaeology*. Cambridge: Cambridge University Press. 104–16.

Tobler, R., Rohrlach, A., Soubrier, J., *et al.* 2017. "Aboriginal mitogenomes reveal 50,000 years of regionalism in Australia." *Nature* 544: 180–4. DOI: 10.1038/nature21416.

Todd, Z. 2014. "Fish pluralities: Human-animal relations and sites of engagement in Paulatuuq, Arctic Canada." *Études/Inuit/Studies* 38(1–2); 217–38. DOI: https://doi.org/10.7202/1028861ar.

Torrence, R. 2011. "Finding the right question: Learning from stone tools on the Willaumez Peninsula, Papua New Guinea." *Archaeology in Oceania* 46(2): 29–41. DOI: 10.1002/j.1834-4453.2011.tb00097.x.

Torrence, R., Kononenko, N., Sheppard, P., *et al.* 2018. "Tattooing tools and the Lapita cultural complex." *Archaeology in Oceania* 53(1): 58–73. DOI: 10.1002/arco.5139.

Townsend, W. H. 1969. "Stone and steel tool use in a New Guinea Society." *Ethnology* 8(2): 199–205. DOI: 10.2307/3772981.

Trigger, B. G. 2006. *A History of Archaeological Thought*. 2nd edn. Cambridge: Cambridge University Press.

Troncoso, A., and Vergara, F. 2013. "History, landscape, and social life: Rock art among hunter-gatherers and farmers in Chile's semi-arid north." *Time and Mind* 6(1): 105–12. DOI: 10.2752/175169713X13518042629379.

Tsing, A. L. 2012. "Unruly edges: Mushrooms as companion species." *Environmental Humanities* 1: 141–54. DOI: 10.1215/22011919-3610012.

Tsing, A. L. 2015. *The Mushroom at the End of the World: On the Possibility of Life in Capitalist Ruins*. Princeton: Princeton University Press.

Tylor, E. B. 1865. *Researches into the Early History of Mankind and the Development of Civilization*: London: n.p.

Vaesen, K., Collard, M., Cosgrove, R., and Roebroeks, W. 2016. "Population size does not explain past changes in cultural complexity." *Proceedings of the National Academy of Sciences* 113(16): E2241–7. DOI: 10.1073/pnas.1520288113.

Valdiosera, C., Günther, T., Vera-Rodríguez, J. C., *et al.* 2018. "Four millennia of Iberian biomolecular prehistory illustrate the impact of prehistoric migrations at the far end of Eurasia." *Proceedings of the National Academy of Sciences* 115(13): 3428–33. DOI: 10.1073/pnas.1717762115.

Van der Leeuw, S. E., and Torrence, R., eds. 1989. *What's New? A Closer Look at the Process of Innovation*. London: Unwin Hyman.

Vander Linden, M. 2007. "For equalities are plural: Reassessing the social in Europe during the third millennium BC." *World Archaeology* 39(2): 177–93. DOI: 10.2307/40026652.

Vander Linden, M. 2011. "A tale of two countries: Contrasting archaeological culture history in British and French archaeology." In B. W. Roberts and M. Vander Linden (eds.), *Investigating Archaeological Cultures: Material Culture, Variability, and Transmission*. London: Springer. 23–40.

Vandkilde, H. 1996. *From Stone to Bronze: The Metalwork of the Late Neolithic and Earliest Bronze Age in Denmark*, Jutland Archaeological Society Publications 32. Aarhus: Aarhus University Press.

Vandkilde, H. 2016. "Bronzization: The Bronze Age as pre-modern globalization." *Praehistorische Zeitschrift* 91(1): 103–23. DOI: 10.1515/pz-2016–0005.

Vavilov, N. I. 1992. *Origin and Geography of Cultivated Plants*. Trans. D. Love. Cambridge: Cambridge University Press.

Venkatesh, V., Morris, M. G., Davis, G. B., and Davis, F. D. 2003. "User acceptance of information technology: Toward a unified view." *MIS Quarterly* 27(3): 425–78. DOI: 10.2307/30036540.

Verbeek, P.-P. 2005. *What Things Do: Philosophical Reflections on Technology, Agency, and Design*. University Park: Pennsylvania State University Press.

Veth, P. 2005. "Cycles of aridity and human mobility: Risk minimization among late Pleistocene foragers of the western desert, Australia." In P. Veth, M. Smith, and P. Hiscock (eds.), *Desert Peoples*. Oxford: Blackwell. 100–15.

Vickers, M. J. 1986. "Silver, copper and ceramics in ancient Athens." In M. J. Vickers (ed.), *Pots and Pans: A Colloquium on Precious Metals and Ceramics in the Muslim, Chinese and Graeco-Roman Worlds, Oxford 1985*. Oxford: Oxford University Press for the Board of the Faculty of Oriental Studies, University of Oxford. 137–51.

Vickers, M. J., and Gill, D. W. J. 1994. *Artful Crafts: Ancient Greek Silverware and Pottery*. Oxford: Clarendon Press.

von Hippel, E. 2005. *Democratizing Innovation*. Cambridge, MA: MIT Press.

Vrydaghs, L., and Denham, T. 2007. "Rethinking agriculture: Introductory thoughts." In T. Denham, J. Iriarte, and L. Vrydaghs (eds.), *Rethinking Agriculture: Archaeological and Ethnoarchaeological Perspectives*. Walnut Creek: CA: Left Coast Press. 1–15.

Wailoo, K., Nelson, A., and Lee, C., eds. 2012. *Genetics and the Unsettled Past: The Collision of DNA, Race, and History*. New Brunswick, NJ: Rutgers University Press.

Wajcman, J. 2000. "Reflection on gender and technology studies: In what state is the art?" *Social Studies of Science* 30(3): 447–64. DOI: 10.1177/030631200030003005.

Wajcman, J. 2009. "Feminist theories of technology." *Cambridge Journal of Economics* 34(1): 143–52. DOI: 10.2307/24232027.

Wallerstein, I. M. 1974. *The Modern World-System*. 3 vols. New York: Academic Press.

Walsh, M. J., Prentiss, A. M., and Riede, F. 2019a. "Introduction to cultural micro-evolutionary research in anthropology and archaeology." In A. M. Prentiss (ed.), *Handbook of Evolutionary Research in Archaeology*. Cham: Springer. 25–48.

Walsh, M. J., Riede, F., and O'Neill, S. 2019b. "Cultural transmission and innovation in archaeology." In A. M. Prentiss (ed.), *Handbook of Evolutionary Research in Archaeology*. Cham: Springer. 49–70.

Warriner, G. K., and Moul, T. M. 1992. "Kinship and personal communication network influences on the adoption of agriculture conservation technology." *Journal of Rural Studies* 8(3): 279–91. DOI: 10.1016/0743-0167(92)90005-Q.

Watson, P. J. 1995. "Archaeology, anthropology, and the culture concept." *American Anthropologist* 97(4): 683–94. DOI: 10.2307/682590.

Wejnert, B. 2002. "Integrating models of diffusion of innovations: A conceptual framework." *Annual Review of Sociology* 28(1): 297–326. DOI: 10.1146/annurev.soc.28.110601.141051.

Wellman, B. 2011. "Is Dunbar's number up?" *British Journal of Psychology* 103(2): 174–6. DOI: 10.1111/j.2044–8295.2011.02075.x.

Wenger, E. 1998. *Communities of Practice: Learning, Meaning, and Identity*. Cambridge: Cambridge University Press.

Wenger, E., McDermott, R. A., and Snyder, W. 2002. *Cultivating Communities of Practice: A Guide to Managing Knowledge*. Boston, MA: Harvard Business School Press.

Wetmore, J. M. 2007. "Amish technology: Reinforcing values and building community." *Technology and Society Magazine, IEEE* 26(2): 10–21. DOI: 10.1109/MTAS.2007.371278.

White, L. A. 1959. *The Evolution of Culture: The Development of Civilization to the Fall of Rome*. New York: McGraw-Hill.

Whiten, A. 2011. "The scope of culture in chimpanzees, humans and ancestral apes." *Philosophical Transactions of the Royal Society: Series B, Biological Sciences* 366(1567): 997–1007. DOI: 10.1098/rstb.2010.0334.

Wilcox, M. 2010. "Marketing conquest and the vanishing Indian: An Indigenous response to Jared Diamond's *Guns, Germs, and Steel* and *Collapse*." *Journal of Social Archaeology* 10(1): 92–117. DOI: 10.1177/1469605309354399.

Wilkins, J. 2018. "The point is the point: Emulative social learning and weapon manufacture in the Middle Stone Age of South Africa." In M. J. O'Brien,

B. Buchanan, and M. Eren (eds.), *Convergent Evolution and Stone Tool Technology*. Cambridge, MA: MIT Press. 153–73.

Wilkins, J. 2020. "Archaeological evidence for human social learning and sociality in the Middle Stone Age of South Africa." In C. Deane-Drummond and A. Fuentes (eds.), *Theology and Evolutionary Anthropology: Dialogues in Wisdom, Humility, and Grace*. London: Routledge.

Williams, A. N., Ulm, S., Turney, C. S. M., Rohde, D., and White, G. 2015. "Holocene demographic changes and the emergence of complex societies in prehistoric Australia." *PLOS ONE* 10(6). DOI: 10.1371/journal.pone.0128661.

Winner, L. 1986. *The Whale and the Reactor: A Search for Limits in an Age of High Technology*. Chicago: University of Chicago Press.

Witmore, C. L. 2007. "Symmetrical archaeology: Excerpts of a manifesto." *World Archaeology* 39(4): 546–62. DOI: 10.2307/40026148.

Wood, S. L., and Moreau, C. P. 2006. "From fear to loathing? How emotion influences the evaluation and early use of innovations." *Journal of Marketing* 70(3): 44–57. DOI: 10.2307/30162100.

Young, A. 2012. "Prehistoric and Romano-British enclosures around the Camel estuary." *Cornish Archaeology* 51: 69–124.

Young, A. 2015. *Lowland Cornwall: The Hidden Landscape*, Vol. V: *Overview*. Truro: Cornwall Council.

Zeder, M. A. 2016. "Domestication as a model system for niche construction theory." *Evolutionary Ecology* 30(2): 325–48. DOI: 10.1007/s10682-015-9801-8.

Zeder, M. A., and Smith, B. D. 2009. "A conversation on agricultural origins: Talking past each other in a crowded room." *Current Anthropology* 50(5): 681–90. DOI: 10.1086/605553.

Zeder, M. A., Bradley, D. G., Emshwiller, E., and Smith, B. D., eds. 2006. *Documenting Domestication: New Genetic and Archaeological Paradigms*. Berkeley: University of California Press.

Zilhão, J. M. 2012. "Personal ornaments and symbolism among the Neanderthals." In S. Elias (ed.), *Developments in Quaternary Sciences*. Amsterdam: Elsevier. 35–49.

Ziman, J. M. 2000. *Technological Innovation as an Evolutionary Process*. Cambridge: Cambridge University Press.

Index